# The Ten Assumptions of Science

# The Ten Assumptions of Science
## Toward a New Scientific Worldview

*Glenn Borchardt, Ph.D.*

iUniverse, Inc.
New York  Lincoln  Shanghai

The Ten Assumptions of Science
Toward a New Scientific Worldview

All Rights Reserved © 2004 by Glenn Borchardt, Director, Progressive Science Institute, Berkeley, California

No part of this book may be reproduced or transmitted in any form or by any means, graphic, electronic, or mechanical, including photocopying, recording, taping, or by any information storage retrieval system, without the written permission of the publisher.

iUniverse, Inc.

For information address:
iUniverse, Inc.
2021 Pine Lake Road, Suite 100
Lincoln, NE 68512
www.iuniverse.com

ISBN: 0-595-31127-X (Pbk)
ISBN: 0-595-66263-3 (Cloth)

Printed in the United States of America

# CONTENTS

PREFACE . . . . . . . . . . . . . . . . . . . . . . . . . . . . . . . . . . . . . . . . . . . . . . ix

INTRODUCTION . . . . . . . . . . . . . . . . . . . . . . . . . . . . . . . . . . . . . . . 1
   Two Views of Metaphysics . . . . . . . . . . . . . . . . . . . . . . . . . . . . . . . 2
      Sense I: Metaphysics is Nonsense . . . . . . . . . . . . . . . . . . . . . . . 2
      Sense II: Metaphysics is the Study of Presuppositions . . . . . . . . 3
   The Nature of Assumptions . . . . . . . . . . . . . . . . . . . . . . . . . . . . . 4
      Differences Between Presuppositions and Assumptions . . . . . . . . 4
      Relativism, Absolutism, and Testability . . . . . . . . . . . . . . . . . . . 4
      Consupponibility . . . . . . . . . . . . . . . . . . . . . . . . . . . . . . . . . . . 6
   Discovering the Assumptions of Science . . . . . . . . . . . . . . . . . . . . 6
      Motivation for the Search . . . . . . . . . . . . . . . . . . . . . . . . . . . . . 6
      Classification of Assumptions . . . . . . . . . . . . . . . . . . . . . . . . . . 7
      Criteria Used for Selecting Assumptions . . . . . . . . . . . . . . . . . . 8
   Outline of the Constellation Used for the Ten Assumptions of Science 9
      Holding and Discarding Assumptions: Dogmatism—Revisionism 11

CHAPTER 1
THE FIRST ASSUMPTION OF SCIENCE: . . . . . . . . . . . . . . . . . . . 15
MATERIALISM . . . . . . . . . . . . . . . . . . . . . . . . . . . . . . . . . . . . . . . 15
   Matter . . . . . . . . . . . . . . . . . . . . . . . . . . . . . . . . . . . . . . . . . . . . 17
   Confirmation . . . . . . . . . . . . . . . . . . . . . . . . . . . . . . . . . . . . . . 19
   Faith and Matter . . . . . . . . . . . . . . . . . . . . . . . . . . . . . . . . . . . . 19

CHAPTER 2
THE SECOND ASSUMPTION OF SCIENCE: . . . . . . . . . . . . . . . . 21
CAUSALITY . . . . . . . . . . . . . . . . . . . . . . . . . . . . . . . . . . . . . . . . . 21
   Specific Causality . . . . . . . . . . . . . . . . . . . . . . . . . . . . . . . . . . . 22
      Absolute Chance . . . . . . . . . . . . . . . . . . . . . . . . . . . . . . . . . . 23
   Finite Universal Causality . . . . . . . . . . . . . . . . . . . . . . . . . . . . . 24
   Infinite Universal Causality (CAUSALITY) . . . . . . . . . . . . . . . . 26
   Example of CAUSALITY . . . . . . . . . . . . . . . . . . . . . . . . . . . . . 27
   CAUSALITY, Motion, and Objectivity . . . . . . . . . . . . . . . . . . . 29

## CHAPTER 3
## THE THIRD ASSUMPTION OF SCIENCE: ...................30
## UNCERTAINTY ............................................30
    The Search for Certainty .................................30
    Determinism: UNCERTAINTY is Subjective ...................31
    Indeterminism: Uncertainty is Objective ....................34
        Is Chance Acausal? ..................................36
        Chance as Ignorance ................................37
        A Dog for an Example ...............................37
        Is Chance a Singular Cause? ..........................40
    UNCERTAINTY and the Unknown .......................42

## CHAPTER 4
## THE FOURTH ASSUMPTION OF SCIENCE: ...............44
## INSEPARABILITY .........................................44
    The Inseparability of Matter and Motion ...................46
    Classical Mechanism ....................................47
        Deterministic Critique ..............................48
        Indeterministic Critique .............................49
    Absolute Zero: Matter Without Motion? ....................50
    Energy: Motion Without Matter? .........................53
    Conceptualizing Matter and Motion .......................55
        Thing-Event ......................................57
        Structure-Function .................................57
        Mass-Velocity ....................................57
        Space-Time ......................................58
    INSEPARABILITY and Clear Thinking .....................60

## CHAPTER 5
## THE FIFTH ASSUMPTION OF SCIENCE: ..................61
## CONSERVATION ..........................................61
    From Atomism to Evolution .............................62
    Challenges to CONSERVATION ..........................64
        Geology .........................................65
        Biology .........................................65
        Cosmology ......................................66
    From the Static to the Dynamic ..........................67

## CHAPTER 6
## THE SIXTH ASSUMPTION OF SCIENCE: ..................68
## COMPLEMENTARITY ........................................68
### The SLT-Order Paradox ....................................68
#### System-Oriented Rationalizations of the Paradox ............70
##### Schroedinger (1967) ......................................71
##### Whyte (1974) ............................................71
##### Makridakis (1977) .......................................72
##### Prigogine (1978) ........................................72
#### Resolution of the Paradox ................................73
### Subjectivity of Order-Disorder .............................75
### Objectivity of Divergence-Convergence .....................76
### The Dialectics of Matter in Motion .........................77

## CHAPTER 7
## THE SEVENTH ASSUMPTION OF SCIENCE: ...............79
## IRREVERSIBILITY ............................................79
### History of IRREVERSIBILITY ...............................80
### How COMPLEMENTARITY Implies IRREVERSIBILITY ......83
#### The Necessity for an Infinite Universe ....................83
### The Myth of Reversibility ..................................84
#### Microscopic "Reversibility" ...............................85
### Does CAUSALITY Require Reversibility? ....................86
#### Time Independence? .....................................86
### IRREVERSIBILITY and the Environment of the System ........87

## CHAPTER 8
## THE EIGHTH ASSUMPTION OF SCIENCE: .................88
## INFINITY ....................................................88
### Macroscopic Infinity and Classical Mechanism ...............89
### Microscopic Infinity and Systems Philosophy ................89
### Quest for the Ultimate Particle .............................92
### Looking for the Edge of the Universe .......................93
### INFINITY: Microscopic Plus Macroscopic ....................93
#### Describing INFINITY .....................................94
#### The Struggle for INFINITY ...............................96

## CHAPTER 9
## THE NINTH ASSUMPTION OF SCIENCE: ............98
### RELATIVISM ............98
#### RELATIVISM Versus Absolutism ............98
#### The Similarity-Dissimilarity Continuum ............100
##### Reasoning by Analogy ............100
##### Reasoning by Disparity ............101
##### Similarity Analysis ............102
##### Examples of the Application of RELATIVISM ............104
###### The Electron ............104
###### The "Conservation" of Parity ............105
#### To Think is to Compare ............105

## CHAPTER 10
## THE TENTH ASSUMPTION OF SCIENCE: ............107
### INTERCONNECTION ............107
#### Disconnection Through the Idea of Perfect Continuity ............109
#### Disconnection Through the Idea of Perfect Discontinuity ............109
#### Search for the Universal Disconnection ............111
##### The Interquantic Interconnection ............113
##### The Intergalactic Interconnection ............114
#### The Necessary Connection ............116
#### The Compleat Indeterminist ............117
#### Interconnections Among the Assumptions ............117
#### Assumptions and the Infinite Universe ............118

## CONCLUSIONS ............119

## INDEX ............121

# PREFACE

As a typical young scientist trained in the United States, I was furnished with wonderful tools for completing my tasks. Not a word, however, was mentioned about the fundamental assumptions I would have to make to carry out the work. Sure, there were inklings about causality and teleology. There were debates with liberal arts students in which we proto-scientists seemed to gravitate toward the side that argued against free will. But an experiment was an experiment; it seemingly did not make any difference what you were thinking when you did it. The curriculum had no room for philosophy, which was then taught as a confusing smorgasbord guaranteed to insult the fewest students and benefactors. Our scientific mentors dimly perceived such "philosophy" to be more hindrance than help.

After all, the key to a successful scientific career seemed to involve sticking to a particular specialty, applying for and getting grants, and avoiding theories outside your field. I could do this well enough, but I kept getting sidetracked. The universe was such an exciting place! The more I studied outside my field, the more suspicious I became about the party line. Imagine! Really smart people were telling us that the whole universe exploded from a point smaller than the period at the end of this sentence. I just couldn't believe it. How could one get serious about such an absurd idea?

As it turned out, the answer was simple: GIGO. Garbage in; garbage out, as the techies say. With the Big Bang Theory, scientists had gotten themselves knee-deep in philosophy and they barely realized it. If nothing else, they had proven once again that it doesn't make any difference how smart you are if you are given the wrong tools to work with. But in science, as in logic, if you don't like your ending point, you need to re-examine your starting point. This book is such a reexamination.

Big Bang theorists, of course, would not agree that a reexamination is necessary. After all, they **like** their ending point. The media and the funders apparently like it even better. Adherents certainly don't think that the Big Bang Theory is "absurd." It really **does** fit their philosophical beliefs, which, as implied in this book, really are not all that scientific. Science has been forced to develop within a world dominated by nonscience. Within the scientific community as well as in the greater society there is a continuous philosophical struggle between "determinism" (the belief that all effects have material causes) and "indeterminism" (the belief that some effects may not have material causes). It is my opinion that, as scientists, we are to be determinists. Part of the philosophical struggle, however, is the production of those who even deny that the struggle occurs or that it is a meaningful activity. They certainly would not be writing a book entitled *The Ten Assumptions of Science*. They are not bothered by the compromises that have produced the current interregnum and its absurdities.

This book covers a lot of philosophical and scientific ground in a short time. I have written it primarily for young scientists and philosophers who will be overthrowing many of the silly theories that my generation has fabricated. I hope that you find it thought provoking and that it benefits your future work immensely. If you find yourself in disagreement with any of these assumptions, I challenge you to come up with your own selections that might be just as consupponible. Who knows? Maybe you can prove that the Big Bang really did occur!

Berkeley, February 26, 2004                               Glenn Borchardt

# INTRODUCTION

*We cannot prove anything except from something that is already admitted.*[1]

In carrying out the first step in science and philosophy—distinguishing one thing from another—strict empiricists of the 19th century collected and classified data without acknowledging an assumptive basis for their activities. In rejecting the religious philosophy that hindered their work, these early scientists found it necessary to reject philosophy altogether. They felt unbiased by preconceived notions. Today this unwarranted righteousness persists even though the collection of data unguided by preconceived notion has been discarded as inefficient, if not impossible. Scientists are expected to test theories and hypotheses. They are to answer questions, not merely collect data.

At the same time, scientists are supposed to be objective. Prejudgment or prejudice is supposed to be the last thing to enter their minds. To admit that preconceived notions or assumptions exist is believed by many to detract from the reputation for objectivity cultivated by scientists everywhere. Thus scientists generally resist the notion that science, like religion, requires an assumptive foundation. In the conventional wisdom, science relies on "facts," while religion relies on "faith."

But facts, like faith, have a way of losing their supposed absolutivity. Facts look different when viewed with new perspectives; perspectives look different when viewed with new facts. It is only when these new facts seem to contradict what we already know to be true that we must reexamine "what we already know to be true," that is, our prejudices or our faith. Indeed, if a scientific faith exists and is an indispensable guide for the scientific method, then the first step in changing that faith is to recognize its existence. Only then can we consciously seek to make it more scientific and thus a better reflection of objective reality. Let us proceed with that first step.

---

[1] Planck, Max. *Where Is Science Going?* London: Allen and Unwin, 1933, p. 195.

## TWO VIEWS OF METAPHYSICS

Because this book is very much about interconnections, I would be remiss in not beginning with a historical connection. In *An Essay on Metaphysics*,[2] the otherwise idealist philosopher R. G. Collingwood made a brilliant attempt to demonstrate the necessary relationship between faith and science. I am going to review this short study in considerable detail because it so clearly demands that assumptive choices must be made. Collingwood implied that both the scientific and religious faiths must be considered as that which goes "beyond physics." Unfortunately, as Collingwood emphasized, Aristotle introduced metaphysics in two fundamentally opposed senses. In the first, the religious or indeterministic sense (Sense I), what is believed to be beyond physics is not physical, but spiritual. In the second, the scientific or deterministic sense, what is believed to be beyond physics is not spiritual, but physical: what is "beyond physics" at any moment is simply more physics (Sense II). From the deterministic perspective, Aristotle's first definition is nonsense, while the second is a major theme of this book. Let us dispense with the nonsense first.

### Sense I: Metaphysics is Nonsense

Aristotle's ever-popular version treated metaphysics as the "science of pure being"—whatever that was supposed to be. But according to Collingwood, a science of pure being would be a science about nothing and is thus a contradiction in terms. There have been innumerable variations on this mystical view of metaphysics in attempts to handle its inherent conceptual and terminological contradictions. It's still a mess.

Stripped of pseudosophistication, metaphysics in Sense I is revealed as gibberish. In Sense I, indeterminists generally posit two realms: the physical and the non-physical. Then they use information necessarily gathered from the physical realm and described in physical terms in an attempt to describe the supposed non-physical realm. The absurdity lies not in extrapolating from the sensed to the not yet sensed, but in the attempt to extrapolate from the sensible to the supposed insensible. Metaphysics in Sense I is truly "nonsense."

It was in reaction to this kind of metaphysics that scientists became antimetaphysicians, with empiricism and its progeny, positivism, being the primary results. For determinist and indeterminist alike, metaphysics was accepted as metaphysics in Sense I. Even dialectical materialists, who were opposed to empiricism and positivism, tended to oppose all metaphysics and any hint that

---

[2] Collingwood, R.G. *An Essay on Metaphysics*. Oxford: Clarendon Press, 1940.

science might legitimately involve faith. If the philosophy of dialectical materialism was metaphysically weak, it was not because it did not contain good metaphysics in some sense other than Sense I, but because it was for the most part unaware that it had any at all. The beauty of Collingwood's essay was that it implicitly recognized the historical confrontation between determinism and indeterminism within metaphysics, rejected most of the indeterministic nonsense, and resurrected the possibility that the deterministic elements of metaphysics may be an essential point of departure for science.

## Sense II: Metaphysics is the Study of Presuppositions

For Collingwood, the second and preferred definition of "metaphysics" was "the science which deals with the presuppositions underlying ordinary science."[3] Like many a historian, Collingwood viewed this science as descriptive rather than prescriptive. His interest was in finding out what presuppositions were made at various points in history, not in suggesting what presuppositions should be made in the future.

This approach potentially legitimizes metaphysics as a science, that is, as a study with an **object** to be studied. Regrettably, Collingwood was more of an idealist than a materialist and his subsequent analyses only managed to support the usual idealistic notions of where presuppositions come from. He carefully avoided any claim that presuppositions were actual physical objects. That idea was left to V. F. Turchin,[4] who astutely regarded scientific concepts as objects undergoing evolutionary regularities similar to those of biological entities. For Turchin, "Knowledge is the presence in the brain of a certain model of reality. An increase in knowledge—the emergence of new models of reality in the brain—is the process of **cognition**."[5]

Thus if we were to suppose for a moment that assumptions exist as material entities present in the brain, we are left with no idealistic mystery as to how they got there. The part played by the interaction of the brain with the external world then cannot be denied. The assumptions in the brain of the priest are just as much material entities as the assumptions in the brain of the astrophysicist. The differences between them arise simply through differing experiences with the external world.

Assumptions, if they are really like other natural objects, cannot remain forever unchanged. Realizing this gives us hope, for our agenda is different from Collingwood's. We study metaphysics, not as an end in itself, but as a means to an end.

---

[3] Ibid., p. 11.

[4] Turchin, V.F. *The Phenomenon of Science*. New York: Columbia University Press, 1977.

[5] Ibid., p. 73.

# THE NATURE OF ASSUMPTIONS

## Differences Between Presuppositions and Assumptions

The highest levels of thinking occur when we realize that thinking begins with presuppositions (e.g., antecedent beliefs unrecognized by the believer). Math states its axioms, logic its premises, and science its assumptions. Maturity is demonstrated here because this way of thinking explicitly acknowledges that other ways of beginning to think about a subject are also possible. Presuppositions rise to the cognitive level of assumptions just as soon as they are discovered, stated, and alternate possibilities recognized.

This analysis assumes that such choices always exist and that presuppositions never stand alone, unopposed. Moreover, the failure to find an alternative for a presupposition does not make it a synthetic *a priori*, just as the failure to find causes for an effect does not mean the effect has no causes. Neither does it mean that thinking could occur without presupposing.

Although there may be any number of alternatives to a particular assumption, striving for clarity tends to reduce the field. The greater the distinction between assumptions, the greater is our confidence that alternate assumptions are true alternates. The greatest disparity between alternate assumptions occurs when they are mutually contradictory, that is, if one is true, then the other is not. The greatest success in stating the fundamental assumptions of a philosophy occurs when we find alternates that negate each other.

## Relativism, Absolutism, and Testability

Collingwood divided presuppositions into two categories: relative and absolute. He believed that the idea of testing applies to the first but not to the second. Relative presuppositions may be "falsified," that is, they may be proven incorrect through an interaction with the external world. Relative presuppositions may be traced back to underlying absolute presuppositions, but the absolute presuppositions can be traced no further. It is at this point that we must differ with Collingwood.

By using the word "absolute," Collingwood implied that such presuppositions exist in miraculous isolation in accord with his idealistic view of where they come from. At this point his astute observation about the historical nature of presuppositions tripped over his absolutist view of what a test is. For Collingwood, a test gives a completely definitive answer to a question (he was not a scientist). Thus relative presuppositions are said to be either completely true or completely false.

It was obvious that his absolute presuppositions could not meet this criterion. Thus he concluded that because absolute presuppositions are not completely testable, not even subconsciously, they are therefore not testable at all. This is a *non sequitur*.

In this way of thinking, the assumption of infinity would be an "absolute presupposition." But every time we find yet another thing in the universe, the assumption of infinity receives validation. The assumption of infinity was derived from our observations of things in the external world, even though it is a contradiction in terms to expect a final test for it. The assumption of infinity did not simply pop into our heads without a material origin.

Collingwood's view was that of the strict logician. He believed that absolute presuppositions were absolute beginning points. They were to be applied to the external world; the external world was not to be applied to them. But this of course contradicts the whole tentative nature of assuming. When we assume, we explicitly admit that other starting points might be possible. We reserve the right to start our thinking all over with a different assumption. Why would we do that? Why would we not continue the chain of logic to wherever it takes us? The answer is that we expect to be dissatisfied with some of the conclusions reached along the way. Strict logicians will not acknowledge the possibility of dissatisfaction because, for them, thinking occurs in isolation. Once the beginning points have been chosen, the deductive thought process enters a closed loop and the messy, infinite complexity of the external world is not allowed to disturb it.

Collingwood's logical perspective not only failed to reveal where absolute presuppositions come from, it also failed to give him an adequate appreciation of the other half of the thinking process: induction. The chain of logic is only a finite model or narrow sampling of the infinite external world. It is of course not the external world itself and thus it inevitably contains statements about the external world that are untrue. All interactions with the external world are tests of our beginning assumptions and none of these tests could provide perfect agreement with our assumptions. It is only when the disagreement is thought to be significant—as in the thinking that produced the Big Bang Theory—that we must go back in search of a new beginning.

The test of a thing is in its use. We shall never know beyond a shred of an indeterminist's doubt whether the universe is infinite or whether all effects have material causes. Nevertheless, we can find out how to reformulate, elaborate, and reinterpret our assumptions so that they will accord more closely with developing experience. No presupposition or assumption could be absolute, isolated, unchangeable, and completely untestable.

## Consupponibility

Similar assumptions, like birds of a feather, tend to flock together. A person who holds a particular assumption is likely to hold other assumptions that are only slightly different. Although one assumption may appear as leader, the unique characteristics of the other assumptions are necessary contributions to the whole. In metaphysics this whole, or group of assumptions, is called a "constellation." A particular constellation of assumptions exists in a unique space-time position; that is, it plays a unique historical role. The relationships among assumptions and between the assumptions and the external world are always in flux. Thus the assumptions underlying our present scientific effort exist in a particular place: in the heads and in the records of people on the planet Earth. They exist at a particular time: the 21$^{st}$ century. Although all constellations will have certain similarities and evolutionary tendencies, no two of them, occurring in different places or at different times, will be perfectly identical.

Any dissimilarity between an assumption and others within a constellation amounts to a contradiction, which if significant, may result in its rejection from the constellation. Any similarity between an assumption and others in a constellation amounts to a confirmation of the suitability of that assumption for membership in the constellation. Assumptions that remain as members of a constellation in spite of their differences are said to be consupponible. In other words, "it must be logically possible for a person who supposes any one of them to suppose concurrently all the rest."[6]

From the indeterministic perspective, whether a particular assumption is to be admitted or retained as part of a constellation would seem to be a purely subjective matter independent of material changes in the external world. From the deterministic perspective this is only partly true, the subjective changes occur, not in spite of the material changes of the external world, but partly because of them. Rapid changes in our material existence must produce rapid changes in the presuppositions and assumptions we require for interaction with the external world.

# DISCOVERING THE ASSUMPTIONS OF SCIENCE

## Motivation for the Search

No one is more interested in the condition of their tools than the tool users themselves. But at times of relaxed vigilance, users have a tendency to delegate

---

[6] Collingwood, *An Essay on Metaphysics*, p. 66.

maintenance to others who may have the specialized training for it, if not the interest. This is what invariably happens soon after a new constellation of assumptions arises through philosophical struggle. The serviceability of a constellation diminishes with time, often unbeknownst to its users. The metaphysicians, those who, according to Collingwood, are supposed to know what the assumptions of science are, tend to lose track of them, much less maintain them. In the period between major changes in constellations, explicit assumptions lose their gloss and fade into implicit presuppositions.

As a rule, the growing inadequacy of a constellation will be noticed first by the scientist, the user, rather than by the metaphysician, the caretaker. The result is that the scientist must become an amateur metaphysician in rebellion against the existing metaphysics. Scientists in this predicament must discover the assumptions of science for themselves, as it is in the nature of crumbling foundations to go for quite some time before their condition or even their existence is acknowledged. The established philosophy that failed to prevent the problem in the first place is not likely to be of much help in this endeavor.

As mentioned, the reason I am so interested in metaphysics and the reason metaphysics is such a large part of this book concerns the fact that I find some of the present-day conclusions of "ordinary science" to be personally unsatisfying. My gut reaction to the notion of massless particles and the claim that the universe exploded out of nothing is one of disbelief. Not being a specialist in these disciplines, I cannot rightly challenge the data used to draw such conclusions, and thus the only way open to me is to trace the interpretations backward to their logical roots. Someone is presupposing something that I am not. Like Collingwood, I would like to know what it is.

The disbelief that initiated this search on my part to challenge current assumptions is also founded upon presuppositions. It takes a lot of faith to be an unbeliever. Anyone who opposes major scientific interpretations must fervently believe that an examination and possible modification of the constellation of assumptions underlying those interpretations can be a potentially fruitful endeavor.

## Classification of Assumptions

Discovering and delimiting the assumptions of science is not easy. The scope of each assumption overlaps others and, depending on personal preferences, one might group some together, split some, or add or subtract others. This work really should be done by a professional, but for the reason given above, such an expectation is unrealistic. Despite the importance of fundamental assumptions, they are seldom made explicit in the scientific literature. Partly, this is because they are taken for granted, being settled, noncontroversial issues until the next scientific revolution.[7] Partly, this is because even a cursory explication requires more space than the average scientific paper allows. Partly, this is because scientific assumptions generally contradict the indeterministic assumptions of the rest of society. Even when deterministic assumptions are advocated, their finer details often are interpreted from the indeterministic point of view.

The title of this book is *The Ten Assumptions of Science*, but the word "the" and the number "ten" are merely expository expedients. The dogmatic use of the word "the" is a way of tweaking those who think that science does not require assumptions at all. As for the "ten," this seems to be a convenient number and appears a rather harmless bow to an indeterministic heritage. This way of cutting up the scientific faith is not the only way it could be done and it will not be the last time it will be done.

## Criteria Used for Selecting Assumptions

The search for assumptions is a search for beginnings. Its success in finding an absolute or ultimate beginning is as likely as finding an origin or a first cause for the universe. If assumptions are disclosed presuppositions, and if presuppositions evolve from other presuppositions in an infinite historical regress, then the criteria used for choosing assumptions will at some point be unknown. The criteria will be presuppositions, not assumptions. Thus the criteria to be used in the selection of assumptions will have themselves evolved from unstated presuppositions. The stated criteria verge on being the assumptions themselves.

I suppose the underlying assumption that forms the criterion for the selection of the other assumptions and for the analysis in this book is that all things in the universe, including ourselves, our actions, and our thoughts are natural consequences of what went before. What many believe to be "free will" is the result of infinitely complicated causes. The path to discovering these causes eventually leads to the external world. I do not agree with Collingwood that the recognizable starting points for science are purely products of the mind. Neither do I agree that they are purely products of the external world, awaiting discovery as self evident, absolute, and unchanging *a priori*. That the historical connections are not always obvious, does not, in my opinion, mean that they never existed. It is of course impossible to completely prove such a proposition because it contains an infinite regress. One can always assert the existence of free will, an absolute starting point, an origin to the universe, or a first cause. I simply choose not to.

---

[7] Kuhn, T.S. *The Structure of Scientific Revolutions*. 2 ed. Chicago: University of Chicago Press, 1970.

If one begins by assuming strict determinism, one must logically end with it. Thus I have chosen only those assumptions and interpretations of assumptions that do not lead to solipsism[8] or fatalism.[9] If I am able to see that a statement proceeds from or leads toward this goal, I call it "deterministic." If not, I call it "indeterministic." The scope of this book requires what will appear to many readers as an extremely large number of such judgment calls. As I said before, the reasons for many of these decisions will be only partly evident, as there is only limited space here for an elaboration that would transform them from the subjective to the objective realm.

In my opinion, the ten assumptions I have selected are consupponible and in agreement with the data of modern science. As mentioned before, a constellation of assumptions is consupponible if it has a low degree of contradiction between members of the constellation. This too, of course, has its subjective element. Diehards can always emphasize the similarities between two assumptions so as to keep both of them within the constellation, while rebels can emphasize the dissimilarities threatening rejection of one or more assumptions.

The decision to confirm or reject would be entirely subjective if assumptions were isolated products of the mind. Because they are not, the accumulating data from the external world impinges upon the subjects, changing their minds concerning what is similar and what is dissimilar. The result is the evolution of the constellation by subtraction, addition, and interaction of assumptions. The subjectivity also extends to the decision on what data are to be admitted as reliable and significant enough to require acceptance or rejection of an assumption. Assumptions and data form a reciprocal relationship.[10] Neither the assumptions nor the data play the dominant role in the decision to consider assumptions consupponible or in accord with the external world.

## OUTLINE OF THE CONSTELLATION USED FOR THE TEN ASSUMPTIONS OF SCIENCE

The Ten Assumptions of Science, the constellation I believe to generally pose the fewest internal as well as external contradictions for the time being, are presented in the next ten chapters. A single capitalized word will refer to these specific assumptions and my interpretations of them. The rejected and opposing indeterministic assumptions will appear in boldface (Table 1).

---

[8] The belief that our internal world completely determines our future.

[9] The belief that our external world completely determines our future.

[10] Sparkes, John. "What Is This Thing Called Science?" *New Scientist* 89 (1981): 156-58.

Table 1. The Ten Assumptions of Science and their opposites.

| No. | *Deterministic Assumption* | *Indeterministic Assumption* |
|---|---|---|
| 1 | MATERIALISM | Immaterialism |
| 2 | CAUSALITY | Acausality |
| 3 | UNCERTAINTY | Certainty |
| 4 | INSEPARABILITY | Separability |
| 5 | CONSERVATION | Creation |
| 6 | COMPLEMENTARITY | Noncomplementarity |
| 7 | IRREVERSIBILITY | Reversibility |
| 8 | INFINITY | Finity |
| 9 | RELATIVISM | Absolutism |
| 10 | INTERCONNECTION | Disconnection |

In brief, the First Assumption of Science, MATERIALISM, posits an external world of material objects that exists after the observer does not. The Second, CAUSALITY, posits an essential connection between material objects so that the motion of one influences the motion of another. The Third, UNCERTAINTY, states that a complete knowledge of an object, cause, or effect is impossible, although an improvement in knowledge is always possible.

The laws of thermodynamics have achieved the status of basic assumptions recognized in all scientific disciplines. The Fourth Assumption of Science, INSEPARABILITY, is a liberally expanded development of the Third Law of Thermodynamics which is, according to this deterministic interpretation, a modern configuration of Hegel's dictum that there can be no motion without matter and no matter without motion. The Fifth, CONSERVATION, the First Law of Thermodynamics, states that matter and the motion of matter can be neither created nor destroyed. The Sixth, COMPLEMENTARITY, asserts the relationship between the Second Law of Thermodynamics and its neomechanical complement, which together, assume that all objects are subject to divergence from and convergence on other objects in the universe. The Seventh, IRREVERSIBILITY, also stems from the deterministic interpretation of the Second Law of Thermodynamics and its complement, which asserts that the interactions of all objects are unique.

The eighth assumption is perhaps the most mind boggling of all assumptions, INFINITY, the proposition that the universe is infinite in both the microscopic and the macroscopic directions. The Ninth, RELATIVISM, assumes that no two things or events are completely similar or completely dissimilar. The Tenth, INTERCONNECTION, assumes that all things in the universe are interconnected and interrelated. This last assumption is illustrated in the assumptions themselves, for they are all interdependent and consupponible to a large degree. An adequate understanding of any one of them requires an understanding of all the others. This is not strictly true for the indeterministic assumptions (Fig. 1).

## Holding and Discarding Assumptions: Dogmatism—Revisionism

Only by ignoring the presupposed nature of its foundation could we deny that science involves an element of opinion rather than **certainty**. Just as there are a potentially infinite number of ways of describing reality, so there are a potentially infinite number of ways of stating the assumptions we recognize as the starting points of those descriptions. I regard the Ten Assumptions of Science as "true" simply because, for me, the weight of observational and experimental evidence that I have considered favors them over others.

As noted previously, no one's assumptions should ever be regarded as isolated notions without any base derived from the real world. Nevertheless, all assumptions and all ideas must be considered somewhat isolated. The real world cannot be experienced *in toto*. Our assumptions must always remain projections of a limited, unique experience. In this sense, assumptions are "metaphysical" because they attempt to go beyond what we have experienced directly. We may assume that the universe is infinite or finite. There is no way for us to prove this beyond a doubt by testing the proposition to its full extent. Still, when we make a choice between such opposing possibilities, certain conclusions follow that have a direct bearing on other interpretations of the world.

Because many people, professional scientists included, believe their own presuppositions and assumptions to be more reasonable than others, they often hold strong opinions about them. They hover about them like mother hens over their baby chicks, protecting them from attack. But like all things, the assumptions of science can be changed, added to, or subtracted from—they evolve. This requires a lot of work. Once the changes have been incorporated within the constellation of assumptions, we tend to relax, claiming *finis* to that which never can be finished. The more faithfully we support a particular constellation, the more dogmatic and unyielding we become. We are likely to be critical, even uncharitable, toward those who contradict us. Dogmatism eliminates the need for a lot of thinking.

12 • The Ten Assumptions of Science

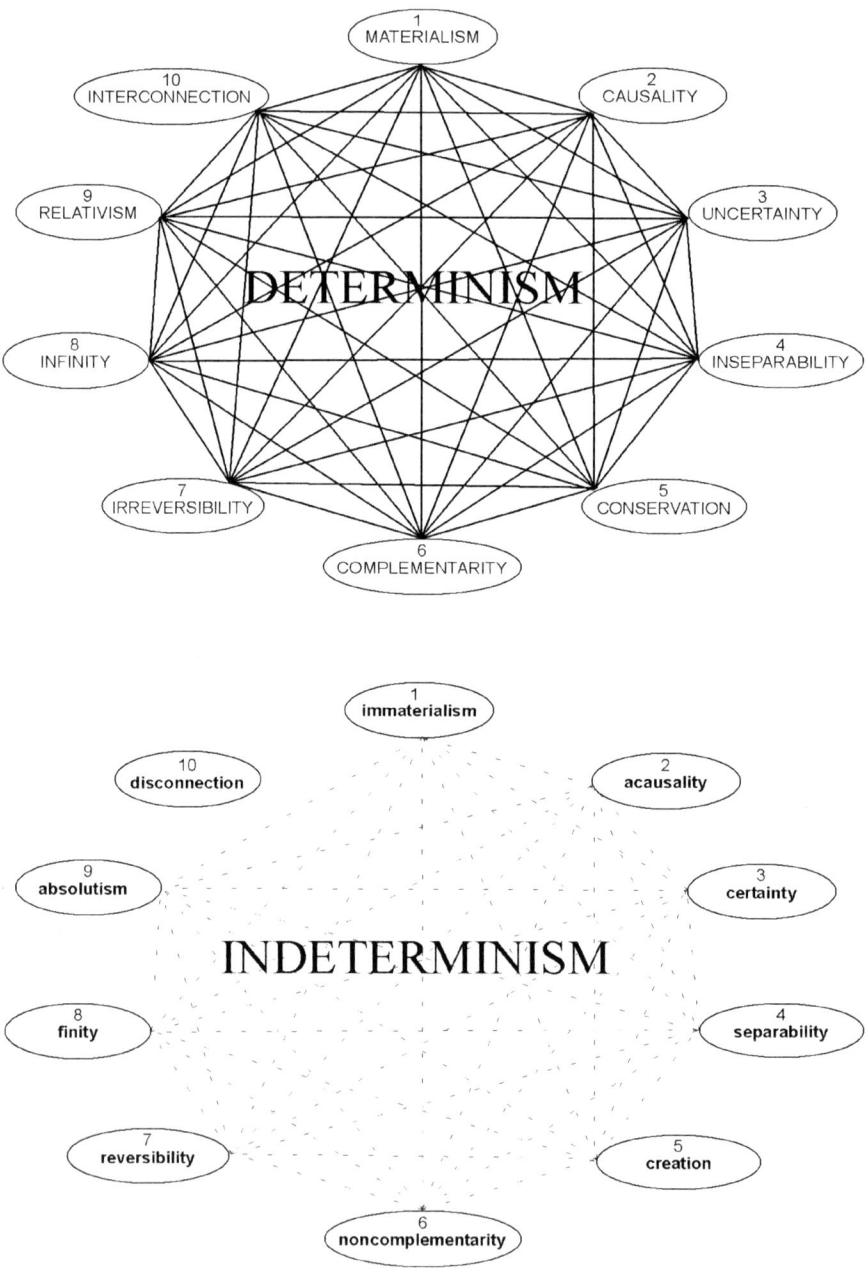

Fig. 1. The Ten Assumptions of Science and the opposing indeterministic constellation. Solid lines indicate the degree of consupponibility.

Supporters of a particular constellation automatically tend toward dogmatism—they naturally believe in preserving assumptions they have found useful. Ideal dogmatists, if they could exist, would be recognized by their steadfast denial that revisions are inevitable. In practice, which revisions are to be admitted and which are not, is a question typically answered in opposing ways by determinists and indeterminists. As we find out more and more about the universe, the necessity for revising our assumptions becomes increasingly pronounced, but always, a relatively unchanged core of belief remains to guide the actions of the most determined iconoclast. In times of philosophical quietude and compromise, we can imagine that this core of belief approaches the shape of a sphere—dogmatists and revisionists become almost indistinguishable. They go almost unnoticed. In times of philosophical evolution and conflict, the imaginary core becomes elongated—dogmatists and revisionists become polarized. Whether an existing dogma or a proposed revision is "good" or "bad," "acceptable" or "unacceptable," is determined, not by logical argument, but by its suitability to the changing conditions of the external world. In the long run, revisionists always win at least partial victories and philosophical progress occurs, but only because the successful changes in dogma are ever more deterministic. In other words, each new dogma results from ever more intensive and extensive interactions of humanity with its environment. The next ten chapters present some daring revisions with profound implications that go far beyond the mere destruction of the Big Bang Theory.

# CHAPTER 1

# THE FIRST ASSUMPTION OF SCIENCE: MATERIALISM

*The external world exists after the observer does not.*

At first thought, MATERIALISM appears obvious. How could anyone believe that the external world does **not** exist? How could anyone **not** be a materialist? Even the etymology of the words "external" and "exists" begs a practical, matter-of-fact acceptance of this, the First Assumption of Science. But as with all Ten Assumptions of Science, experience can provide only **support** for MATERIALISM; it cannot **prove** it beyond a shred of an indeterminist's doubt.

There will **always** be otherwise intelligent beings who mistakenly believe that the existence of the external world is somehow dependent on them. Thus in the heat of philosophical conflict it often becomes necessary to state the obvious. Even the most celebrated scientist of the 20th century, Albert Einstein, was moved to remind himself and others that "The belief in an external world independent of the perceiving subject is the basis of all natural science."[11] This slightly ambiguous definition of MATERIALISM nevertheless expresses a commandment of scientific faith. If we did not assume this to be true, doubts could creep into our analysis at the most crucial moments.

Through the ages not a few indeterministic philosophers have become rich and famous trying to prove just the opposite: **immaterialism,** the assumption that reality is internally derived and that the external world exists at most as a series of sense impressions. The latter point of view reached its zenith in the philosophy known as **subjective idealism**. In typical fashion its major proponent, Bishop Berkeley, was led to write:

---

[11] Einstein, Albert. *Ideas and Opinions.* New York: Crown, 1954, p. 266.

> It is indeed an opinion strangely prevailing amongst men, that houses, mountains, rivers and in a word all sensible objects, have an existence...distinct from their being perceived by the understanding. But this principle involves a manifest contradiction. For what are the aforementioned objects but the things we perceive by sense? ...Their *esse'* is *percipi*; nor is it possible they should have any existence out of the minds which perceive them.[12]

Today, this extremely solipsistic view is usually confined to animals such as ostriches and to infants, although subtle remnants of it exist in all of us. The Berkeleian past echoes in the sentiments of those who ask: "If a tree falls in the forest and there is no one there to hear it, does it still make a sound?" Naive realists,[13] as well as most practicing scientists, quickly say, yes it does. But idealists, preferring the subjective definition of sound, are not so sure.

One might think that scientists, at any rate, would be in unanimous agreement that only MATERIALISM bears the test of experience. This is not the case, for **immaterialism** has always had numerous backers within the scientific community. For instance, in summing up the enigmatic achievements of 20$^{th}$ century physics, one of its principle spokesmen, Professor John Wheeler of the University of Texas had this to say in 1979:

> What is so hard is to give up thinking of nature as a machine that goes on independent of the observer. What we conceive of as reality...is but the elaborate work of our imagination. For our picture of the world, this is the most revolutionary thing discovered. We have still not come to terms with it.[14]

At the cutting edge of science, where it is likely to do the most good, Wheeler urges the abandonment of MATERIALISM in favor of its indeterministic alternative. This tendency to waver on the belief in MATERIALISM appears to be an occupational disease among theoretical physicists, especially since the end of the 19$^{th}$ century.

Even though the majority of scientists will continue to ignore Wheeler's outworn suggestion, there are many subtle ways of giving an indeterministic cast to MATERIALISM. For example, in the Einstein definition, it is not completely

---

[12] Berkeley, George. "Mind and Matter." In *Introductory Readings in Philosophy*, edited by M.G. Singer and R.R. Ammerman. New York: Charles Scribner's Sons, 1710 [1962].

[13] A person, unschooled in philosophy, who believes that the external world closely resembles one's perception of it.

[14] Begley, Sharon. "Causality, King Chance." *Newsweek*, March 12 1979.

clear in what way the observer should be seen as independent of the object observed. This is a question of "epistemology," the study of knowing, which will be discussed in more detail later.

A common indeterministic interpretation is an extension of the naive realism that produced MATERIALISM in the first place. Naive realists tend to believe that the objects of the external world are self evident and directly perceived, as they exist in reality. Their picture of the external world comes in loud and clear and there are no transmission problems. In asserting the physical independence between the observer and the object observed, they often tend to deny the material interconnections necessary for perception.

We observe the external world, however, only through our five senses—far from perfect instruments. The naive realist is likely to be enamored with sensory perception in the visible part of the spectrum, forgetting, or being unaware that this is only a tiny fraction of the radiation emitted from material objects. We cannot see x-ray or infrared radiation, but these also are interactions occurring in the material world. The "reality" one would derive from vision in these parts of the electromagnetic spectrum would be considerably different from the one we know. Being ourselves finite, material portions of an infinitely rich universe, we are limited. We see the world as it appears; it is impossible to see it as it really is.

Scientists, too, sometimes claim more objectivity than they are capable of. These claims often can be traced to the naive interpretation of MATERIALISM. Of course, the separation of subject from object is never completely successful and the goal of pure objectivity is never achieved. Just as the object must interact with the subject to produce an assumption in the first place, the subject, in turn, must interact with the object to apply the assumption.

The definition of MATERIALISM that I prefer (**the external world exists after the observer does not**) attempts to avoid the possibility of an indeterministic epistemological interpretation by emphasizing the metaphysical nature of the assumption. This definition certainly does not admit of a complete, ultimately personal test. There will never be a way for me to know for certain that the universe exists after I do not. Nevertheless the indirect and incomplete evidence that is available to me is sufficient to lead me to the "leap of faith" that is the primary basis for all science. Personally, I do not consider this "leap" to be of any consequence, but I see no reason to deny that it exists.

## MATTER

According to MATERIALISM, the sensory impressions we receive from the external world result from the motions of matter. "Matter" is defined as an abstraction for the world of physical objects. It is the name we give to the class that includes all things. The category "matter" is like the category "fruit." One cannot eat a "fruit," one can only eat a specific kind of fruit such as an apple or an orange. Thus in the strictest sense, matter, like fruit, does not exist; only specific, concrete examples of matter can exist. Matter is definitely not what it was once thought to be: a sort of filling or substance within the indivisible atoms of which the external world is composed.

In the modern view, matter consists of specific objects within specific objects *ad infinitum*. Even though conceptions of the nature of matter vary considerably, those emphasizing its importance usually assume it continues to exist after the observer is gone.

This objective view stresses the importance of the external world of matter, and thus is philosophically opposed to subjective idealism. And because it emphasizes the importance of matter, this philosophy is called materialism. Materialism is "a theory that physical matter is the only or fundamental reality and that all being and processes and phenomena can be explained as manifestations or results of matter."[15] A materialist might be heard to say: "You and I can **imagine** spirits, ghosts, and all sorts of things, but when I hit this wall with my hand, it's **real!**" Science is, by definition, materialistic in this way, demanding that its theories and explanations confront the external world.

Without these confrontations, without observations and experiments, and without a firm belief in MATERIALISM, science may slip into idealism, the belief in **immaterialism**, its opposite. As we have seen, this is what happens from time to time in disciplines in which the confrontation with the external world is especially difficult to achieve. Scientific or intellectual work that is done in relative isolation from the outside world invariably suffers from a degree of subjective idealism. It is only when that relative isolation is broken that the merits of MATERIALISM are reaffirmed. Eventually the theoretician is put back on course by the experimentalist.

Whether or not we admit it, we are all materialists. We derive from the external world the "faith," "assumption," or "knowledge" that the external world exists. We could not walk another step if we did not. But the characteristics of matter are not everywhere the same. We may imagine that the ground ahead will support our weight, or we may imagine that it will not. Determinists simply claim that the way to find out is to interact with that portion of the external world. As Max Planck, the famous physicist put it: "The chief quality to be

---

[15] *Webster's New Collegiate Dictionary*. Springfield, MA: Merriam, 1979.

looked for in the physicist's worldpicture must be the closest possible accord between the real world and the world of sensory experience."[16]

## CONFIRMATION

Anyone who has suffered impaired sensation or has experienced hallucinations knows how difficult it is to distinguish the external world from the internal world. Sometimes, to make this distinction, we need a little help from our friends. We assume that matter has certain characteristics that others can sense in ways similar to our own. To improve our confidence in our sensory perceptions, we seek confirmation of our initial conclusions. We do this either by repeating our observations and experiments, or by comparing them with those of others.

This desire for confirmation makes science a group effort. Even the most independent scientists rely on thousand of volumes of scientific reports and accumulated data. The words of departed scientists echo from the past, challenging those who follow to confirm or to reject their interpretations. All scientists must make their work available for confirmation or give up pretensions of advancing knowledge. Those who see some feature of the world in a special way must be able to teach others how to see it too. By interacting with others we discover our failures in perception and adjust our viewpoint to be more in tune with reality. The collective "faith" of scientists differs from the collective "faith" of religious believers only in so far as it reflects a more intensive and extensive experience with the external world.

Confirmation is sought in all walks of life. People avidly seek testimony from others who support their opinions. When this support is not forthcoming, people become disappointed, censorious—even violent. If contradiction brings war, confirmation brings peace. Ironically, those who feel contradicted by the assumption of MATERIALISM often behave in ways that betray their need for confirmation and their doubts that they can exist in the world without interacting with it.

## FAITH AND MATTER

Except for a few theoreticians in modern physics, most scientists consider MATERIALISM hardly worthy of debate. Nevertheless, within science, indeterministic interpretations of MATERIALISM stem from naive realism or from nonfeuerbachian interpretations of "faith." Both conflict with the deterministic viewpoint because, like overt **immaterialism** they hypothesize a physical disconnection

---

[16] Planck, *Where Is Science Going?* p. 85.

between humans and their surroundings that simply does not exist. Over 160 years ago Ludwig Feuerbach[17] showed that faiths of all kinds were derived from the material existence of the people that held them. The gods of warriors were warriors; the gods of shepherds were shepherds. Contrary to the believers in "free will," a faith, whether deemed religious or scientific, does not simply pop into one's head out of nowhere. The validity or truth of a particular faith, assumption, or bit of knowledge cannot be decided merely by imagining that it was caused or uncaused, but by testing it in the external world.

---

[17] Feuerbach, Ludwig. *The Essence of Christianity*. New York: Harper Torchbooks, 1841.

# CHAPTER 2

# THE SECOND ASSUMPTION OF SCIENCE:
# CAUSALITY

*All effects have an infinite number of material causes.*

The First Assumption of Science posits an external world consisting of material objects, but it does not necessarily require one to believe that those objects exist in anything but rigidly fixed positions, each isolated from the other. Indeed there have been materialists who have tried to maintain such a static view of matter as a way of avoiding the Second Assumption of Science, CAUSALITY.

The concept of causality assumes that the objects of the external world are in motion relative to each other and that all objects are influenced by the motions of other objects. Without a belief in causality we could still believe that things exist outside ourselves, and we might observe them, but we could not link their motions; we could make no interpretations or predictions. The motions of objects would appear nonsensical, and life would be a meaningless blur of events. In the opinion of Hans Reichenbach, the famous positivist[18] philosopher of science, "if we did not believe in causality, there would be no science."[19] In science, as in life, we seek meaning by discovering the causes of effects.

---

[18] Like empiricists, positivists contend that sense perceptions are the only admissible basis of human knowledge and that assumptions are unnecessary for describing the phenomena that we experience. Positivists hold that it is impossible to say anything about phenomena not yet observed. Thus, for example, a strict positivist would never hypothesize the existence of an atomic particle or an earthquake fault for which there is no direct evidence.

[19] Reichenbach, Hans. *The Rise of Scientific Philosophy*. Berkeley: University of California Press, 1951, p. 42. With this passage, Reichenbach betrays the positivist doctrine, finding it necessary to discard his belief in no belief.

But as we shall see from Reichenbach himself, there are widely varying interpretations of the notions of causality and its opposite, acausality. If causality is the proposition that **all** objects are influenced by the motions of other objects, then **acausality** is the absurd proposition that **no** objects are influenced by the motions of other objects. With only a few exceptions associated with the philosophy of modern physics,[20,21] most indeterminists do not advocate this extreme position. Instead of invoking **acausality** as a generalization, they invoke it only in specific instances. To most people at least a few events seem to have obvious material causes. Nowadays it would be difficult, for instance, to convince educated people that the motion of the leaves waving in the wind was not caused by the influence of moving air.

As I see it, there are three major views of causality: "specific causality," which assumes that some effects have material causes, but that others may not; "finite universal causality," which assumes that all effects have a finite number of material causes; and "infinite universal causality," which assumes that all effects have an infinite number of material causes. In this chapter and the next I show that only infinite universal causality should be considered the modern, deterministic interpretation of CAUSALITY. The others, however, must be understood as well.

# SPECIFIC CAUSALITY

As numerous devout scientists have shown, one does not require a belief in a generalized version of causality to do specialized work. Moreover, in the strictest sense one doesn't even need to assume a cause exists to look for one: "We can leave this question open, like the question of what is the cause. Only if we knew that there is no cause would it be unreasonable to seek for a particular cause."[22]

In dealing with specific problems, the indeterminist believes rightly that general or universal causality "is certainly not the logical presupposition of the particular causal law under investigation."[23] This pragmatic position is all that is required to begin scientific investigations.

For the indeterminist, the major advantage of specific causality is its consupponibility or agreement with its opposite: specific acausality. Some objects may appear to be influenced by the motion of other objects, but other objects appear to be uninfluenced by the motion of other objects. By maintaining that

---

[20] Russell, Bertrand. *Mysticism and Logic*. New York: Longmans Green, 1918.
[21] ———. *On the Philosophy of Science*. New York: Bobbs-Merrill, 1965, p. 163.
[22] Reichenbach, *The Rise of Scientific Philosophy*, p. 112.
[23] Ibid., p. 113.

the external world is only partly dynamic, materialists could avoid being strict determinists.

Unfortunately, the association of causality with motion and acausality with the lack of motion has proven to be an errant mistress for the indeterminist. Ever since Einstein disposed of the possibility of a preferred frame of reference, thus confirming the view introduced by Heraclitus that "everything is in flux," the notion of absolute rest has taken a beating. If no thing is perfectly static, then no event is acausal.

## Absolute Chance

If specific acausality and indeterminism were to be rescued from the ruins created by a dynamic external world, the association between motion and causality had to be soft-pedaled. Early in the history of philosophy Aristotle developed the possibilities for a lasting divorce. As Aristotle saw it, events come about in three ways: 1) By external compulsion, 2) By internal compulsion, or 3) Without definite causes but by absolute chance.[24] C. S. Peirce considered the doctrine of absolute chance to be the "utmost essence of Aristotelianism."[25] Indeed, it deserves to be called the essence of 20th century philosophy and science. Wherever the doctrine of absolute chance is invoked, the association of causality with motion is severed cleanly. Like his descendents, Aristotle did not deny causality altogether—that would have been intellectual suicide. Instead, he assumed causality for specific instances and denied it for others. If one were to make an assumption of universal causality, the contradiction with its opposite, universal acausality, would become obvious and would force a choice between determinism and indeterminism. The doctrine of absolute chance neatly avoided that.

The disjunction between causality and motion allowed Aristotle and his followers to return to the *argumentum ad ignorantiam*: because none can be found, therefore the cause for a particular effect does not exist. This dogma naturally led to the association of causality with law and absolute chance with lawlessness. Whatever could be described by means of causal laws was causal; whatever could not was not. For some, this version of specific causality had an added virtue: compatibility with **immaterialism**. Even as a sometime believer in causality, Reichenbach, in particular, was moved to resurrect Berkeley's ghost: "The relations controlling unobserved objects violate the postulates of causality."[26]

---

[24] Peirce, C.S. "The Doctrine of Necessity Examined." In *Determinism, Free Will, and Moral Responsibility*, edited by Gerald Dworkin, 33-48. Englewood Cliffs, NJ: Prentice-Hall, 1970, p. 33.

[25] Ibid., p. 41.

[26] Reichenbach, *The Rise of Scientific Philosophy*, p. 183.

It is often difficult to know when a specific law or assumption is applicable. It is appropriate, however, that this subjective interpretation of causality chooses as its limitation, the boundary between the observed and the unobserved. Thus according to Reichenbach: "Causality is an empirical law and holds only for macroscopic objects, whereas it breaks down in the atomic domain."[27] But as it turns out, the events of the atomic domain are not completely unobservable. Some information can be obtained, although it is indirect and generally has a great deal of statistical variation. The nature of the probability distributions drawn from such data will be discussed in the next chapter. For now, it may suffice to note the similarity between Aristotle's view of absolute chance and Reichenbach's 20th century view of the laws of probability: "The idea of strict causality is to be abandoned and the laws of probability take over the place once occupied by the law of causality."[28]

There are varying interpretations of what is being said here, but, as I will argue in more detail later, there is no fundamental difference between the notion of absolute chance and this view of the laws of probability. Both amount to an assumption of specific acausality developed as a logical complement to the assumption of specific causality. Their veiled philosophical purpose is to avoid an assumption of universal causality.

Despite its great success in specialized science, we cannot use specific causality as the basis for scientific philosophy. The failure of specific causality is demonstrated whenever one attempts to use it for making general philosophical and scientific statements. This is perhaps the most flagrant abuse of specific causality, because even a vague familiarity with logic should remind us that a deduction cannot be more general than its premise.

## FINITE UNIVERSAL CAUSALITY

In spite of Aristotle's great influence on 20th century science, the most broadly experienced scientists still insisted on employing a universal form of belief in causality. On occasion, such sentiments are made explicit even in reports on specialized topics. For example, in a paper discussing the origin of life, Linus Pauling, the twice-named Nobel laureate, and his coauthor, Emile Zuckerkandl, reminded their readers: "Causality, determinism—this rule is considered to apply

---

[27] Ibid., p. 307.
[28] Ibid., p. 163.

intrinsically, to the relation between **all** phenomena, until proof to the contrary is forthcoming."[29]

In one form or another, the extreme, generalized view of causality persists in spite of the indeterministic borrowings from ancient Greece. Even staunch indeterminists must admit: "Science has advanced in the past precisely because, when things happened whose causes were unknown, it was assumed that they had causes nevertheless."[30]

The evolution of the belief in the applicability of the universal generalization was a tremendous advance in thinking—one that is now more easily accepted than 2400 years ago when Democritus first introduced it. It took living organisms billions of years to transcend the myopia of specific causality. Today, we can do it in a fraction of a lifetime. Generalists and interdisciplinary scientists often develop a belief in universal causality after first using specific causality in narrow disciplines. Subsequently, and often subconsciously, they find that a new, specific assumption for each investigation is unnecessary. Unless they are continually admonished, they naturally slip into the carefree habit of assuming universal causality.

There are two major forms of universal causality: finite universal causality and infinite universal causality. Let us review finite universal causality first, since it appeared first and remains the most common conception of universal causality.

Perhaps the best explanation of finite universal causality was give by Pierre Simon Laplace, the philosopher-scientist who, independently of Kant, advanced the nebular hypothesis of the origin of the solar system. Laplace illustrated his view of determinism by hypothesizing a super intelligent being that has come to be known as Laplace's Demon:

> We ought to regard the present state of the universe as the effect of its antecedent state and as the cause of the state that is to follow. An intelligence, who for a given instant should be acquainted with all the forces by which nature is animated, and with the several positions of the beings composing it, if his intellect were vast enough to submit these data to analysis, would include in one and the same formula the movement of the largest bodies in the universe and those of the lightest

---

[29] Pauling, Linus, and Emile Zuckerkandl. "Chance in Evolution: Some Philosophical Remarks." Edited by D.L. Rohlfing and A.I. Oparin, *Molecular Evolution: Prebiological and Biological.* New York: Plenum Press, 1972, p. 120.

[30] Blandshard, Brand. "The Case for Determinism." In *Determinism and Freedom in the Modern Age of Science: A Philosophical Symposium*, edited by Sidney Hook. New York: New York University Press, 1958, p. 9.

atom. Nothing would be uncertain for him, the future as well as the past would be present to his eyes."[31]

As did Einstein, a few old-fashioned "determinists" still hold to this view although it has suffered at the hands of determinists and indeterminists alike. We now recognize that Laplacian determinism is invalid because it contradicts a major Assumption of Science, INFINITY, to which Einstein, of course, did not subscribe. In his fanciful illustration, Laplace was implying that the cause of a particular effect could be determined with absolutely perfect precision, that the motion of a particular body is determined solely by a finite number of the motions of other bodies.

But any concept of knowledge also requires the concept of subject and object. In 1927 Werner Heisenberg presented the Uncertainty Principle, which demonstrated that the knowledge required of some objects, at least, could not be obtained without interfering with those objects. The interference produces changes in motion that, in turn, cannot be evaluated without additional interference with the object. This leads to an infinite progression in which, theoretically, Laplace's Demon would require infinite time to determine the position and momentum of a single object. The demon would be so busy in this effort, that it would be forced to ignore the rest of the universe. Unobtrusively, the assumption of INFINITY, the materialist theory of knowledge, and the Heisenberg Uncertainty Principle presided over the death of Laplacian determinism and the theory of finite universal causality.

## INFINITE UNIVERSAL CAUSALITY (CAUSALITY)

After Heisenberg's discovery, the hopes for a consistent determinism grew dim. But thirty years later, David Bohm, the philosopher-physicist, showed that still another view of universal causality was possible. In his masterful classic, *Causality and Chance in Modern Physics*,[32] Bohm presented an especially elegant exposition that inspired at least one author to consider him the first to prove "the logical possibility of a deterministic model."[33] Bohm showed that the quantum mechanical laws to which the Heisenberg Uncertainty Principle applies could not be assumed to be inviolate. As for statistical probability in general, Bohm pointed

---

[31] Quoted in Castell, Alburey. *An Introduction to Modern Philosophy*. 3 ed. New York: Macmillan, 1976, p. 520.

[32] Bohm, David. *Causality and Chance in Modern Physics*. New York: Harper and Brothers, 1957.

[33] Hawkins, David. "The Thermodynamics of Purpose." In *Beyond the Edge of Certainty*, edited by R.G. Colodny, 102-17. Englewood Cliffs, NJ: Prentice-Hall, 1965, p. 116.

out that future investigations of the atom are likely to uncover causal laws that explain some of this present uncertainty. Even so, the infinity of nature will always require a statistical approach each time we arrive at one of these deeper levels of organization. Although we temporarily may be unable to find a material cause for a particular phenomenon, the cause nevertheless exists.

Bohm's explanation of universal causality contains a more or less explicit assumption of infinity. He points out, for example, that the "cause" for an "effect" is never established with absolute certainty.[34] We must always accept something less because both the cause and the effect, like the objects they describe, have infinite properties. Nevertheless, some of these properties are more important than others, and by concentrating our attention on the most important, we define a cause for an effect. Like all good pragmatists, we are satisfied with the approximation as long as it is useful to us. At a later date we may find that additional factors are important in predicting the effects of causes and so, we will include them in our considerations. In no case, however, should we delude ourselves into thinking we have discovered all of the factors involved, for in principle, they are infinite in number.

# EXAMPLE OF CAUSALITY

As an illustration of the assumption of infinite universal causality (CAUSALITY) let us consider the position of the Earth in relation to the sun. Newton's law of gravitation states that the gravitational effect between two objects increases with mass and decreases with distance. At a particular moment, this law provides an approximation of the distance between the Earth and the sun. The predicted distance and the actual distance, however, are never identical. This is because the gravitational fields of other astronomical bodies also influence both the Earth and the sun. The moon, for instance, causes a noticeable wobble in the path described by the Earth around the sun. So do other planets, but their effects are only a couple percent of the moon's effect. By including the effects of the gravitational fields of all the planets in addition to that of the moon, we can develop a finite set of equations that will predict the distance between the Earth and the sun with reasonable accuracy. The prediction, however, can never be **perfectly** accurate. There always will be astronomical bodies in addition to the ones we have considered that will contribute gravitational effects upon the Earth. These influences are so slight that we can usually neglect them.

---

[34] Bohm, *Causality and Chance in Modern Physics*, p. 20 and p. 22.

In practice, we **must** neglect some of them, because the number of astronomical bodies is so great—100 billion stars in each of 100 billion galaxies estimated to be within the view of our telescopes alone.

The achievements of Newtonian celestial mechanics arose out of such judicious neglect and amounted to the reduction of the infinite complexity of the universe to the simple consideration of just a few bodies. The practical, mathematical, necessity of working with a finite number of causes engendered the belief in finite universal causality and its corollary, the idealistic belief in the possibility of perfect accuracy. Although this approach was, in the main, highly successful, it always had a disconcerting element of failure. Each claim for perfect accuracy was eventually overthrown by more detailed work showing that prediction and reality did not coincide exactly. Closer agreement could be obtained only by expanding the number of causes used to predict a particular effect. CAUSALITY assumes that, because the universe is macroscopically and microscopically infinite, the number of these causes is in principle infinite. Practical success is achieved in the way the mechanists did it: by reducing the number of causes to those having the most significant impact on the problem at hand. The philosophical distinction between the finite and the infinite versions of causality is thus simply this: CAUSALITY consciously assumes that failure in prediction at some point is inevitable, and always requires a search for additional or more appropriate material causes for its correction.

The analysis so far presented in this example is enough to establish that a phenomenon as "simple" as the distance between the Earth and the sun is affected by the gravitational effects of a theoretically infinite number of astronomical bodies, all of which contribute to the whole, but most of which contribute very little. In addition to this static analysis we can superimpose the fact that all of those astronomical bodies are in motion relative to each other. According to RELATIVISM (to be discussed in Chapter 9) neither the distance between any two bodies nor the mass of either of them is exactly the same at any two moments. By the time we have determined the distances and masses of whatever bodies we are going to use in predicting the distance between the Earth and the sun, the actual values will have changed and so will the actual distance between the Earth and the sun. As a result, no two trips of the Earth around the sun describe exactly the same path or have exactly the same duration.

The overconfidence sometimes engendered by Laplacian determinism, Newtonian mechanism, and their reductionist approach is unwarranted. We can never give a **complete** description or perfectly accurate prediction of any phenomenon because, as we assume under INFINITY, the number of objects to be considered is infinite and each of them is in motion relative to each other and all other bodies in the universe. In practice we may discover a finite number of causes that will enable us to describe and predict with great accuracy and precision, but this should not delude us into believing that the number of causes for a particular effect is finite.

## CAUSALITY, MOTION, AND OBJECTIVITY

The assumption of infinite universal causality hereafter referred to as the assumption of CAUSALITY, reunites the concept of causality with the concept of motion. CAUSALITY is dynamic and objective, not static and subjective. MATERIALISM posits an external world of matter, while CAUSALITY posits the dynamic interaction between its parts. Neither its unending complexity, nor its unending motion allows us to give a complete causal statement of the motions of even one object relative to all other objects. Matter in motion simply will not sit still for it. Nevertheless, we are able to compile partial, finite statements that we call causal laws and that we find relatively valid for specific instances. An auxiliary belief in uncaused motion can be only a hindrance in this effort.

# CHAPTER 3

# THE THIRD ASSUMPTION OF SCIENCE:
# UNCERTAINTY

*It is impossible to know everything about anything, but it is always possible to know more about anything.*

According to Pliny the Elder, nothing is certain but uncertainty. Individuals, their philosophies, and their scientific endeavors suffer whenever they cannot handle uncertainty. To be uncertain is one of the most uncomfortable of feelings. Life seems to be a never-ending search for certainty, punctuated by premature announcements that it has been found at last. For millennia, people have looked toward philosophy for the comfort of absolute certainty. The Third Assumption of Science, however, states that they will find only UNCERTAINTY instead.

While CAUSALITY is a statement about the interactions between objects, UNCERTAINTY is a statement about what may be discovered about those interactions. For us, the relationship between subject and object is as important as the relationship between objects. To understand CAUSALITY, we must also understand UNCERTAINTY. To fail to do this is to remain a captive of $20^{th}$ century philosophies, best characterized by their enslavement to Aristotelianism. As I will show in this chapter, the philosophical choice that we must make is clear: either causality is objective and uncertainty is subjective, or causality is subjective and uncertainty is objective. The historical roots of the way in which scientists make this choice lie within the longstanding search for certainty.

## THE SEARCH FOR CERTAINTY

In many ways the search for certainty and the search for knowledge are the same, but there are two ways of viewing that search. Indeterminists traditionally have approached the quest with the idea that absolute certainty or absolute knowledge actually could be found. Determinists, associated with the rise of scientific philosophy, have given up the search for the absolute and have accepted a relative instead. We no longer seek perfect accuracy and omniscience, only better accuracy and more knowledge. With the advent of the Heisenberg Uncertainty Principle, philosophers and scientists were forced to take a new look at the causality-uncertainty relationship.

When Heisenberg[35] demonstrated that it was impossible to know **both** the momentum **and** the position of a particle at the same time, he hit a sore spot, especially with the Laplacian "determinists." Essentially, he was telling them that, at least in the world of subatomic particles, science was "limited" and aspects of it were uncertain. If one could not be certain of both the momentum and the position of a particle through time, then one could not be absolutely certain of the relation between cause and effect either. Bertrand Russell, the philosopher-mathematician, captured the disillusionment of that period: "The hope of finding perfection and finality and certainty has been lost."[36] Science was forced to admit that CAUSALITY and UNCERTAINTY were indubitably linked and would have to be **assumed**; there could be no absolute certainty. One could never know everything about anything.

The search for absolute certainty nevertheless continued. For many, the Heisenberg Uncertainty Principle took on its own kind of certainty. Mystical physicists spoke of it as an opportunity to "dethrone the law of causation."[37] Many were sure that uncertainty meant there was a "loose jointedness" that destroyed the argument for absolutely strict causation. Russell declared, "the reason why physics has ceased to look for causes is that, in fact, there are no such things."[38] For some, the rationale for indeterminism and its implications for the doctrine of free will brightened.

## DETERMINISM: UNCERTAINTY IS SUBJECTIVE

According to Hermann von Helmholtz, it was possible to deduce all phenomena from their causes in a logically strict and uniquely determined manner. The search for certainty led scientists of the 19<sup>th</sup> century to equate the orderly

---

[35] Heisenberg, W. von. "Uber Den Anschaulichen Inhalt Der Quantentheoretischen Kinematik Und Mechanik." *Zeitschrift fu"r Physik* 43 (1927): 172-98.

[36] Quoted in Lakatos, Imre, John Worrall, and Gregory Currie. *Mathematics, Science and Epistemology*. Vol. 2. New York: Cambridge University Press, 1978, p. 18.

[37] Jeans, Sir J.H. *The Mysterious Universe*. New York: Macmillan, 1930, p. 17.

[38] Russell, *On the Philosophy of Science*, p. 163.

motions of the external world with the orderly motions of the internal world. Natural law and humanly devised law were thought to be identical.

With Laplace and Newton, "determinism" had lapsed into indeterminism; it demanded perfection where perfection was not possible. There may be causes for all events, but it was pure idealism to believe that anyone could know all the causes for even one event. If, as Bohm had proposed, causality involved INFINITY, then the old view of determinism as both objective and subjective had to be discarded.

In deciding between these alternatives it is instructive to consider the etymology of the word "determinism." Like the word "law," it originally described a uniquely human activity. But with the growing belief in MATERIALISM, both "determinism" and "law" began to take on an objective meaning as well. The objects of the external world were said to "determine" each other independent of a perceiving subject. In the Newtonian view, the motion of one body was "deterred" or "terminated" by that of another. Although there always have been some who saw natural law in a teleological sense, there were others who saw it in a strictly objective sense. Whatever those objects did to each other occurred because there were no other possibilities under the conditions and not because the objects were following a predetermined script.

Einstein, Planck, and De Broglie remained stoutly opposed to the growing tendency of physicists to interpret the Heisenberg Uncertainty Principle as a sign of acausality in nature. Einstein, in particular, was vociferous in mounting this opposition from the point of view of traditional, Laplacian determinism. The argument failed because, the stronger the case was made for finite universal causality, the greater was the tendency to associate determinism with finite, humanly devised laws. If the temporarily unknown portions of the external world eventually had yielded completely to this mathematically based treatment, then the laws of the scientist and the laws of nature would have coincided perfectly, as Einstein had hoped. Nature, however, refused to cooperate. Being infinite in number, and always in motion relative to each other, the various parts of the external world produced phenomena and natural laws faster than anyone could describe them. In the last analysis causality could not be proven to be both objective and subjective.

In forcing a choice between the objective and the subjective views, the Heisenberg Uncertainty Principle unintentionally laid the groundwork for a new form of determinism. As David Bohm put it:

> None of the really well-founded conclusions that can be obtained with the aid of the assumption of a finite number of qualities in nature can possibly be lost if we assume instead that the number of such qualities

is infinite, and at the same time recognize the role of contexts, conditions, and degrees of approximation. All that we can lose is the illusion that we have good grounds for supposing that in principle we can, or eventually will be able, to predict everything that exists in the universe in every context and under all possible conditions.[39]

With this single statement Bohm nicely demonstrated the consupponibility of MATERIALISM, CAUSALITY, and INFINITY while implying that UNCERTAINTY is strictly a statement about the subject-object relationship. With this, Bohm systematically destroyed the case for Laplacian determinism, not from the indeterministic perspective characteristic of Heisenberg and the "Copenhagen" school, but from a revolutionarily new deterministic view. He rejected entirely Einstein's dream of a single equation that would describe the fundamental characteristics of the universe, allow perfectly accurate prediction, and thus lead to absolute certainty. As Bohm pointed out, "there is no real case known of a set of **perfect** one-to-one causal relationships that could in principle make possible predictions of **unlimited** precision..."[40] Finally, "we do not expect that any causal relationships will represent **absolute truths**; for to do this, they would have to apply without **approximation**, and unconditionally."[41]

UNCERTAINTY states that no matter how good our measurement of real objects, an improvement in precision is always possible, although perfect precision is not. There is always yet another cause that will explain part of the variation, but because the number of causes is infinite, some variation, some uncertainty will still remain.

The consequence for practical science is that no relationship between real objects is strictly linear.[42] That is, the plot of one real variable against another real variable really cannot give a perfectly straight line. In simple mathematical terms, when we say that some result or effect, z, is a function of causes x and y (i.e., $z = f(x,y)$), what we really mean is that: $z = f(x,y,...\infty)$. When we are asked: "What is the cause of the variation of z?" We say "factors x and y," ignoring the infinite number of other factors because we feel they are unimportant for practical purposes.

---

[39] Bohm, *Causality and Chance in Modern Physics*, p. 135.
[40] Ibid., p. 20.
[41] Ibid., p. 32.
[42] Weinberg, G.M. *An Introduction to General Systems Thinking*. New York: Wiley-Interscience, 1975, p. 232.

As Bohm reiterated, "every real causal relationship, which necessarily operates in a finite context, has been found to be subject to contingencies arising outside the context in question."[43]

We can always add a finite number of these causes or conditions to an expanded study, but obviously we cannot add an infinite number of them. As a result, "science is never in a position completely and exhaustively to explain the problems it has to face…the solution of one problem only unveils the mystery of another. We must accept this as a hard-and-fast irrefutable fact."[44]

And as Weinberg put it: "If we want to learn anything, we mustn't try to learn everything."[45] We must focus on the main features of an object and its environment. We can always include additional features for consideration, but we can never include all features, for they are infinite in number. Thus, UNCERTAINTY is correctly expressed as a subject-object relationship: **It is impossible to know everything about anything, but it is always possible to know more about anything.**

## INDETERMINISM: UNCERTAINTY IS OBJECTIVE

For modern Aristotelians who believe that some events may not have material causes, there is only one way in which to view uncertainty: as an inherent characteristic of object-object interactions. Thus according to this, the so-called "Copenhagen" view, we must "consider indeterminacy in the conditions of material bodies as an objective state of affairs."[46] To maintain logical consistency, the flip side of this particular indeterministic coin requires the belief that "this indeterminacy is not caused by the limitation of our mental horizon to some specific segment of the world."[47] According to the Copenhagen school, the uncertainty concept has nothing to do with how we see the world. It is not a problem of our trying to squeeze infinite information into finite equations. The failure of Laplacian determinism is not "because our descriptions are in need of correction, but because there is always in nature a certain…indeterminacy."[48]

---

[43] Bohm, *Causality and Chance in Modern Physics*, p. 3.

[44] Planck, *Where Is Science Going?* p. 82-3.

[45] Weinberg, *An Introduction to General Systems Thinking*, p. 105.

[46] Elsasser, W.M. *Atom and Organism*. Princeton, NJ: Princeton University Press, 1966, p. 89.

[47] Ibid., p. 78.

[48] Collingwood, R.G. *The Idea of Nature*. Oxford: Clarendon Press, 1945. pp. 123-4.

Today both the determinist and the indeterminist pretty much agree that the perfectly accurate predictions promised by Laplace, Newton, and Einstein are impossible. The limitations of trying to view the infinite universe from a single point inevitably intrude upon the CAUSALITY-UNCERTAINTY debate. Either uncertainty is objective and causality subjective or *vice versa*. The indeterminist urges us to "formulate the principle of causality in all strictness as a proposition concerning cognitions, instead of trying to understand it as one concerning things and events."[49]

This subjective view of causality is really **immaterialism** in disguise. Occasionally it ends up stark naked, as in the comment by Reichenbach that "the relations controlling unobserved objects violate the postulates of causality,"[50] or in the one by J. D. Fast that "there is nothing in nature to which a definite position **and** definite velocity can be simultaneously attributed, because no particles, in the classical meaning of the word, exist."[51]

At the least, the subjective view of causality denies the connection between motion and causality. At the most, it denies the existence of the very objects to which the motion refers.

As mentioned in the previous chapter, subjective causality is essentially specific causality. By definition, whatever is included in the humanly constructed causal law is causal; whatever is not included is acausal. While extremists might be satisfied with that, moderates generally prefer other terminology. They realize that, at least in the macroscopic world, additional investigations often lead to the development of causal laws for phenomena formerly considered acausal. Historically, this has been the way in which the indeterministic position has been eroded. Thus indeterminists have sought ways in which to put an end to the dissipation of their claims. The Copenhagen interpretation of quantum mechanics seemed to reach such an endpoint, becoming a major contributor to the resurgence of indeterminism in the 20th century.

On the other hand, if there is to be a renaissance of determinism, the Copenhagen interpretation must be discarded. The choice to be made is becoming clearer. The deterministic position, as long as it includes Bohm's concept of infinite universal causality, remains unchanged with the development of knowledge.

---

[49] Cassirer, Ernst. *Determinism and Indeterminism in Modern Physics: Historical and Systematic Studies of the Problem of Causality.* New Haven, CT: Yale University Press, 1956, p. 65.
[50] Reichenbach, *The Rise of Scientific Philosophy*, p. 183.
[51] Fast, J.D. Entropy: *The Significance of the Concept of Entropy and Its Applications in Science and Technology.* New York: McGraw-Hill, 1962, p. 174.

The indeterministic position on uncertainty as an objective state of affairs, however, yields to every increase in knowledge. What is at first regarded as uncertain and acausal later may be regarded as certain and causal. The recent development of chaos theory is a case in point.[52]

To demonstrate further that the deterministic view of UNCERTAINTY is the correct one, I will attempt to answer two questions: Is chance acausal? Is chance a singular cause? A partial answer to the first question is elementary to every investigation. In the simple example to follow I will demonstrate that, at least in the macroscopic world, chance can be considered just another word for ignorance. The second question concerns whether such an analysis can be extrapolated into the microscopic world—the subject of the philosophy surrounding the Heisenberg Uncertainty Principle. Supporters of the Copenhagen view do not see chance in quantum mechanics as a matter of ignorance and many of them do not regard it as acausal. On the contrary, they tend to view chance as a singular cause. Clues to the development of this widely held belief reveal its kinship with Laplacian determinism, the limitations of mathematics, and the abiding search for **certainty**.

## Is Chance Acausal?

In the preferred definition, "chance" is "something that happens unpredictably without discernible human intention or observable cause."[53] In indeterministic definitions the words "human" and "observable" are conveniently forgotten in the effort to promote the unknown and unobserved as a sign of acausality in the universe. In this regard, Aristotle became the greatest of all sinners when he appended the word "absolute" to the word "chance." For most indeterminists, it makes little difference anyhow, because they link both absolute chance and regular chance with the notion of acausality. When indeterminists speak of "luck," "fortune," "fate," or "chance," they tend to forget that there are causes for the events described by these words. Chance may be viewed in one of three ways:

1. As observer ignorance,
2. As a sign of acausality, or
3. As a singular cause.

It is relatively simple to show that the first is more plausible than the second.

---

[52] Scheck, Florian. *Mechanics: From Newton's Laws to Deterministic Chaos.* Berlin; New York: Springer-Verlag, 1999.
[53] Webster's New Collegiate Dictionary, 1979.

## Chance as Ignorance

A major purpose of education is to eliminate as much guesswork as possible. The more we know about a system and its relationship to its environment, the less we need to use chance in describing it. In the parlance of scientific research, what we know rather definitely may be described by deterministic models, while what we know only vaguely must be described by probabilistic models. The laws of probability used in developing probabilistic models are merely tools for delimiting our ignorance. As our knowledge increases about a thing or an event, as we include more and more of the infinite number of causes of certain effects, the knowledge component of the relationship increases and the ignorance component decreases. But ignorance always remains. Some variation—some uncertainty—always exists in every prediction. Indeterminists can forever point at this uncertainty and claim it as an indication of acausality, but determinists, accustomed to removing successive layers of ignorance and of developing knowledge from what once was mystery, will not be impressed.

## A Dog for an Example

The conversion of the "probabilistic" into the "deterministic" through investigation can be illustrated by a commonplace example. Virtually all probabilistic models use a variant of the normal distribution or "bell-shaped curve" for describing a property of a particular class of objects (Fig. 3-1). In our hypothetical example, animals classed as dogs have been weighed, their weights have been plotted on the horizontal scale, and the frequencies of the various weights have been plotted on the vertical scale. The result is a bell-shaped curve that tells us that dogs weigh about 15 kg on the average and that their weights range from less than a kilogram to more than 30 kg. Such general information, of course, is of extremely limited usefulness in making certain predictions. For instance, it would not give us a very precise estimate of the weight of the next dog we will see on the street. About all we can say is that there is a greater likelihood that this dog will weigh about 15 kg than, say, 3 kg or 40 kg. Our state of ignorance about what it will be like is so great that we must use statistics, probability, or "chance" in discussing its weight.

People who consider the dog as a "system," isolated from all else, most likely would throw up their hands and remark something like "Your guess is as good as mine." At the most, they would be forced to use the entire bell-shaped curve (Fig. 3-1) in making their estimate. Logically, they would bet that the next dog seen on the street would be of average weight—something around 10 to 20 kg. Those more scientifically inclined, however, would increase the odds in their favor by recalling that dogs are really not isolated from their environments. By knowing more about how a dog's size relates to its environment we can eliminate some of the "chance" indicated by the large standard deviation in Fig. 3-1.

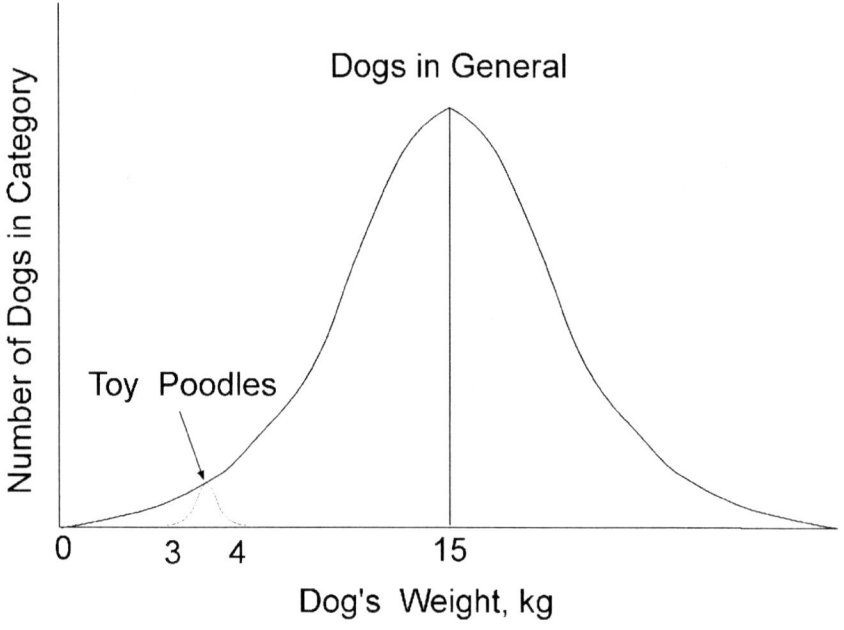

Fig. 3-1. Hypothetical distribution of dogs' weights.

Suppose that we talk with someone on the street and find that a neighborhood is noted for its annual "Toy Poodle Show." We might infer that there is a good reason for the event being held in this particular location. A nonscientist might see absolutely no possibility of a relationship between annual Toy Poodle Shows and the sizes of local dogs. After all, some people travel hundreds of miles to dog shows. Scientists, however, play such guessing games all the time. They realize that, more often than not, related systems exist nearby rather than far away. Human beings and their cultural paraphernalia follow the same pattern.[54]

We know that toy poodles are smaller than most dogs and we also know that the variation in weight for adult dogs is less than for dogs as a whole. If we investigate further and find out that the apartments in the area allow only "lap" dogs, then this information further reduces the element of chance in our guess

---

[54] Zipf, G.K. Human Behavior and the Principle of Least Effort: An Introduction to Human Ecology. Cambridge, MA: Addison-Wesley, 1949.

about the weight of the next dog we will see on the street. Adult toy poodles have weights between 3 and 4 kg. By weighing a large number of them we could get another bell-shaped curve. Of course the width of the bell will be tiny compared to the one for dogs in general (Fig. 3-1).

In this way, more and more "chance" is removed as we learn more and more about the dog whose weight we are about to guess. When the dog actually appears and we see that it is a male poodle and that he appears fat instead of thin, we have eliminated even more "chance." Next, we might remember that it is 10:15 A.M., the time when Mrs. Smith, who is very punctual, takes her grand champion poodle out for a walk. And oh yes, last year the grand champion at the "Toy Poodle Show" weighed.... We could continue on and on, decreasing ignorance and "chance" with every step, but never eliminating it altogether.

The process illustrated above is typical of the scientific method and the development of all knowledge. Based on a limited amount of knowledge, we guess at the possible relationships between things, subject our guess to further observation or experiment, and then accept or reject the guess for use in making still more guesses. This view of chance as a subject-object relationship also helps us understand the nature of knowledge itself. Knowledge and ignorance must be seen as relatives—knowledge is nothing without ignorance.

We may congratulate ourselves for being on the scene at exactly 10:15 A.M. and for having guessed the dog's weight in the above example, but the guess could not be perfectly accurate. Even if our estimate is within 0.1 kg, there is still infinite detail to be discovered in that last 0.1 kg. We would find that a dog's weight fluctuates with such normally trivial factors as temperature, relative humidity, and the growth and activities of parasites and microorganisms. Finally, a living creature's weight fluctuates with each breath that it takes. Minutia such as these should dispel any notion that a dog's weight is the result of a finite number of causes.

No matter what level of detail we reach, the infinite character of the world produces an uncertainty in every analysis. Each data point and each "constant" determined from real objects and their motions has a plus and minus associated with it. A good analyst knows that expending further effort to eliminate major causes of variation can reduce this variation. But all variation cannot be eliminated, for two reasons: 1) increasing effort is required to determine and to remove ever more insignificant causes of variation, and 2) the system itself as well as its environment is in constant motion—a fact that became inescapable in Heisenberg's consideration of the microscopic realm. During the period in which the causes of some of the variation are being discovered, the system and its environment changes in ways that present new variations requiring further discovery. According to UNCERTAINTY, there can be no system—macroscopic or microscopic—that does not necessitate a continual updating of our knowledge about it.

## Is Chance a Singular Cause?

The Heisenberg Uncertainty Principle was destined to trigger a revolution in physics, but its immediate effect was a counterrevolution. The concept of finite universal causality that it destroyed was deeply ingrained. As always, scientists continued their practical search for certainty—for the complete description. As with Einstein, the belief in its ultimate success continued as a matter of habit. Thus Bohm's theory of infinite universal causality met with marginal acceptance. Mathematics, promoted as the language of mature science, would never develop a dialect of infinite length. Rather than adopt an assumption that defied mathematical treatment, the sophisticated members of the Copenhagen school took a step backward—toward Laplace. Regarding chance as a singular cause could preserve the illusion of certainty.

This was advantageous to those who shunned an overt association with the notion of acausality. It satisfied the demand for completeness and was consistent with the atomistic idea that the quantum was indivisible. Through the special pleas engineered by the Copenhagen school, the microscopic realm could be considered a contravention of UNCERTAINTY. And even if Bohm's hypothetical "subquantic states" were eventually discovered, the argument still could be used against the inevitable future hypothesis of "subsubquantic states."

The notion of chance as singular cause grew out of the reductionistic practice of lumping the probabilistic aspects of an analysis within a single category. Instead of the Bohmian equation: $z = f(x,y,...\infty)$, we get the neo-Laplacian equation: $z = f(x,y,C)$, where C includes all of the infinite number of poorly known factors under a single category referred to as "chance." The property, $z$, is then regarded as a result of only three factors, $x$, $y$, and $C$, in much the same way that the Laplacian determinists had insisted. In narrative the concept of chance as singular cause appears like this: "The origin of a body like the Earth depends exclusively upon chance plus the properties of the elements...and the other known factors."[55]

Whether one accepts such a description as complete depends on one's view of the concept of chance. If chance is viewed as we view it in the assumption of UNCERTAINTY, then it is multiple instead of singular. Only in the crudest sense is a description containing "chance" complete. As long as further investigation can yield additional knowledge regarding the causal factors lumped under "chance," a description or theory is not complete.

---

[55] Henderson, Lawrence Joseph. *The Fitness of the Environment; an Inquiry into the Biological Significance of the Properties of Matter.* New York: The Macmillan Company, 1913, p. 304.

The Heisenberg Uncertainty Principle and quantum theory provided a grand opportunity for indeterminists to promote the Aristotelian notion of chance as singular cause and once again announce the successful conclusion of the search for **certainty**. Theorists of the Copenhagen school felt sure that, at least in the microscopic realm, a nonamendable theory had been found. Construed as a singular cause, the concept of chance did what indeterminists had always wanted to do: call a halt to scientific activity—the establishment of cause and effect.

Still, there was the standard deterministic objection: "To assume that a frustration of present knowledge, even one that looks permanent, is a sign of chance in nature is both practically uncourageous and theoretically a *non sequitur*."[56]

True enough, but at the time that the Copenhagen school made its greatest inroads, physics was fast becoming, as some would say, "overly" mathematized.[57] "Practice" in modern physics increasingly became mathematical practice rather than experimental practice. The demands of mathematics for finitude produced, in the heads of $20^{th}$ century physicists, a finite world far removed from the relentless infinitude of reality.[58]

The common sense gained through "hands-on" experience with the macroscopic realm gave way to a plethora of equations about the microscopic realm. With its mechanistic assumption of **finity** the Copenhagen interpretation of the Uncertainty Principle claimed to do what the classical mechanists had always wanted to do but could not: eliminate the admission of ignorance from explanation. With waveforms and probability distributions, physicists perfumed the garbage dump of the unknown and resurrected Aristotle's absolute chance.

Whether considered acausal or singular cause, chance was once again credited with mystical powers. In the name of chance, the public has had to endure outrageous fantasies masquerading as science. Sir Arthur Eddington, ever on the lookout for an opportunity to spread indeterminism, could seriously propose the idea that objects would have a certain remote possibility of disappearing into thin air simply because the atoms within them undergo motions ascribed to "chance."[59]

Continuing in this vein, Eddington promulgated the ridiculous notion that, given enough time, and enough monkeys and typewriters, the monkeys would eventually type all the great books. By "chance" they would eventually hit all the keys in the correct sequence. These bizarre extrapolations invariably emphasized

---

[56] Blandshard, "The Case for Determinism," p. 9.

[57] Honig, W. "Mathematics in Physical Science, or Why the Tail Wags the Dog." *Speculations in Science and Technology* 2 (1979): 361-62.

[58] Edmonds, J.D., Jr. "Can a World without Infinity Be Compatible with the Real Numbers?" *Speculations in Science and Technology* 1 (1978): 79-83.

[59] Blandshard, "The Case for Determinism," p. 9.

calculations showing astronomical, but never impossible odds. Their philosophical purpose was to divert attention from the real conditions under which real events happen. The mathematics of the probability distribution in Fig. 3-1 allows for the existence of a dog weighing several tons, but we would be foolish to go looking for one.

Throughout the 20th century, even serious scientific explanations resorted to the concept of chance as singular cause. A classic example concerns the origin of life on Earth. Reichenbach once declared that "the laws of probability...eventually produce higher and higher forms of life."[60] Jacob Bronowski, the great popularizer of science, reiterated the same antiquated view: "The manifestations of life...must contain a large element of the accidental."[61] Similarly, the normally perceptive Carl Sagan, wrote that "The evolution of life on Earth is a product of random events, chance mutations, and individually unlikely steps."[62] But chance, being no more than observer ignorance, can produce nothing at all. There are reasons for each so-called "accident" or "random event" that make it highly likely—indeed, the only possibility under the conditions. Specialists involved in such studies believe that "not only is any hypothesis based on the chance occurrence of rare events not subject to experimental test, but also that such hypotheses are contrary to most of the available evidence."[63]

# UNCERTAINTY AND THE UNKNOWN

With the assumption of UNCERTAINTY, scientific philosophy shows its potential for developing a refreshing and liberating view of the universe. While it necessarily must be somewhat dogmatic in holding its assumptions, it need not be absolutely dogmatic. While it continually searches for certainty and for knowledge, it need not claim that absolute certainty and absolute knowledge can be found. The universe is far too complex for claims that any part of it could be described in full or its motions predicted in complete detail. Wherever we discontinue our describing or theorizing, whether it be at the point where indeterminists would call the rest of what we do not know "chance," or at the

---

[60] Reichenbach, *The Rise of Scientific Philosophy*, p. 200.

[61] Bronowski, Jacob. *A Sense of the Future: Essays in Natural Philosophy*. Cambridge, MA: MIT Press, 1977, p. 291.

[62] Sagan, Carl. *The Cosmic Connection: An Extraterrestrial Perspective*. New York: Dell, 1973, p. 43.

[63] Kenyon, D.H., and Gary Steinman. *Biochemical Predestination*. New York: McGraw-Hill, 1969, p. 31.

point where exhaustion overcomes us, let us admit our ignorance. The potential for knowledge is infinite.

Every theory has within it a time bomb that eventually destroys it—the Laplacian "assumption that all physical factors (mass and/or energy) which enter into a reaction are known, that all possible parameters have been defined."[64] The Aristotelian belief in chance as singular cause only gives one a false sense of **certainty**. It invariably associates chance with knowledge rather than with ignorance. The question of the nature of chance is basic to an understanding of science. It will continue to arise again and again in the pages to follow. In all of science, a special effort must be made to avoid using words such as "chance," "accident," "random," or "luck" as indications of anything other than observer ignorance. To do so would be a violation of the Third Assumption of Science, UNCERTAINTY.

---

[64] Dudley, H.C. "Is There an Ether?" *Industrial Research*, November 15 1974, 41-46, p. 41.

# CHAPTER 4

# THE FOURTH ASSUMPTION OF SCIENCE:
# INSEPARABILITY

*Just as there is no motion without matter, so there is no matter without motion.*[65]

In effect, MATERIALISM posits matter and CAUSALITY posits motion. The Fourth Assumption of Science declares the essential connection between the two concepts. Like MATERIALISM and CAUSALITY, INSEPARABILITY seems mere common sense. How could anyone speak of motion without acknowledging the existence of the thing that moved? How could anyone, at this late date, speak of motionless matter? But the opposing assumption, **separability** does just that. It is one of the venerable underpinnings of indeterminism.

The idea that **all** portions of the universe are in motion relative to every other portion would have been inconceivable in primitive society. Indeed, to the primitive, most objects appeared motionless and those that did not, seemed to move as though by some supernatural power. The presupposition that later gave rise to the indeterministic assumption of **separability** exploited an unavoidable conceptual difficulty. The results were as fantastic as the physical connection between matter and motion was difficult to prove. From **separability** we get our rich heritage of immaterial "things" and "beings" nevertheless presumed capable of motion. Without **separability**, souls, ghosts, and gods would find no place in the universe. Even within science, the residuum of related ideas persists. Aided by the impossibility of conceptualizing both matter and motion at the same time, **separability** seems to give everlasting life to indeterminism.

From the very beginning, the belief in determinism was inextricably associated with the belief in the connection between matter and motion. From Democritus in the 5th century B.C. to Spinoza and Hobbes in the 17th, deterministic philosophers

---

[65] Hegel, G.W.F.

presented the view that the existence of matter and the occurrence of its motion relative to other matter was natural, not supernatural.

As always, the more intensive and varied one's experience with the material world, the more doubtful are indeterministic notions, particularly of **separability**. The Industrial Revolution, in particular, stimulated the desire and the necessity for a "hands-on" approach to understanding things and their movements. With the development of machines came the need for a general theory of their operation. This was met in 1687 when Isaac Newton published the *Principia*, thereby laying the theoretical and mathematical foundation for the science of mechanics. Being much concerned with objects and their motions, this new methodology unavoidably spread the newly evolving notion of INSEPARABILITY. The practical success of mechanics, even in fields removed from the industrial arts, promoted the first reasonably consistent version of the scientific worldview: mechanism. "Mechanism" took its major cue from INSEPARABILITY and advanced the revolutionary conviction that the universe consists solely of matter in motion.

None of this occurred without an intense struggle. Indeterminists found mechanism easy to criticize. The term itself conjured images of noisy, dehumanizing contraptions in a polluted atmosphere of smoke, grease, and oil. Sophists regarded the reduction of the infinite complexity of the world to two conceptual categories as hopelessly crude. Nevertheless, this idea, born of such lowly parentage, swept the intellectual world, becoming the guiding light of science until the 20$^{th}$ century. From the motions of the planets to the manifestations of consciousness, nothing was too complicated for the explanations of a Holbach, Büchner, or Loeb, who promoted mechanism and its underlying, stillborn assumption of INSEPARABILITY.

But as mentioned in the previous two chapters, the reign of the old-time mechanists ended with the death of Laplacian determinism. Classical mechanism was hoist on its own petard. To support their claims that complete explanations were possible, mechanists had to concede an end to the motion they began with. It was impossible to give a perfectly complete description of systems that changed even before the describing was done. Instead of admitting this, mechanists tended to revert to atomism, the belief that objects are immutable and therefore completely describable and entirely predictable. In supporting finite universal causality to the bitter end, classical mechanism ended up denying the protoevolutionary idea that gave rise to its birth.

Unfortunately, the overthrow of mechanism has placed INSEPARABILITY itself in doubt. During the transition from a Laplacian to a Bohmian variety of determinism we seem to have lost our way. Once again, we must suffer tales of matter without motion and of motion without matter. In addition, we are confronted with a new twist: the claim that matter **is** motion and that motion **is** matter. To find the scientific path again, we must disinter INSEPARABILITY.

## THE INSEPARABILITY OF MATTER AND MOTION

If any topic can be said to be ultimately unfathomable, it is the "relationship" between matter and motion. In the strict sense we cannot even speak of such a relationship because relationships occur only between "things," portions of the universe. As I argue in this section, motion is not a "thing," despite what we must say about "it." Movement cannot be a "part" of the universe because it is the activity of parts of the universe, not the parts themselves. It is really not legitimate to ask, as some have,[66] how matter and motion are connected, because that would be treating motion as a thing, as matter, which it is not. The unity of the world consists of there not being a **physical** distinction between matter and motion,[67] even though its correct explication necessarily requires a **conceptual** distinction.[68] The philosophical issue cannot arise unless one insists on a physical as well as a conceptual distinction as indeterminists are wont to do.[69]

The philosophical issue was supposedly put to rest with the equation that made Einstein famous: $E = mc^2$. Defined here as the mathematical product of matter and motion, the concept of energy was to guarantee physical INSEPARABILITY for all time. But has it? How many people really understand that the conceptual unification that Einstein was trying to achieve is, in the end, impossible? Einstein's mathematization of the energy concept could not and did not prevent the conceptual vulgarization of matter-motion to **either** matter **or** motion.

The main thesis of this chapter is that to understand properly the physical INSEPARABILITY of matter and motion we must regard them as conceptually distinct. As will become clearer after the chapter on INFINITY, we are able to do this in a way that Einstein was not. His belief in finity led to the conceptual and mathematical closure that gave him the equation, and along with it several mystifications for the enjoyment of indeterminists everywhere.

The naive realist, on the contrary, intuitively knows that there is a big difference between matter and motion. The phenomenon of "legs" and the phenomenon of "running" cannot be conceived as identical no matter how much we

---

[66] Gregory, Frederick. *Scientific Materialism in Nineteenth Century Germany*. Boston: Reidel, 1977, p. 157.

[67] Collingwood, *The Idea of Nature*, p. 23 and p. 167.

[68] Gregory, *Scientific Materialism*, p. 107.

[69] Buchner, Ludwig. *Force and Matter*. 15 ed. New York: Truth Seeker, 1884, p. 30.

try. The common sense despised so much by modern physics nevertheless cannot consider "running" to be material in the same sense that "legs" are considered material. "Running" does not have existence in the sense that "legs" have existence. One can possess legs, but one cannot possess running. Running is what legs do; motion is what matter does.

Under MATERIALISM, we assumed that the universe consisted of matter, and that all things, being portions of the universe, were examples of matter. All portions of the universe have dimensions and therefore have existence. Strictly speaking then, motion *per se* does not exist, only things, only matter can exist. There is no place left in the universe where motion could exist. What then, **is** motion if it is not a thing, does not have existence, and is not independent of matter? As mentioned, motion is what matter does. We know, as Einstein emphasized, that each portion of the universe continually changes position relative to all other portions of the universe. The fact that this occurs is as sure as the fact that matter exists.

Furthermore, neither matter nor motion should be considered more important than the other; both are "primary." Motion cannot occur without matter and matter cannot exist without motion. Or, as Hegel put it: **Just as there is no motion without matter, so there is no matter without motion**.

While the first part of Hegel's dictum is relatively easy to understand, the second part is not. For many people the deduction does not follow from the premise. But if one agrees with Engels that "matter is unthinkable without motion,"[70] then Hegel's statement is the way in which INSEPARABILITY must be understood. What gives an object its materiality is, first, its consisting of other objects in **motion**, and second, its existing among other objects in **motion**. As we shall see later, when we inquire, in turn, after these objects, we get the same answer. The question "What is matter?" goes begging into INFINITY. In deterministic usage, motion always refers to an object that is moving relative to other objects that may or may not be explicitly mentioned. Again, the existence of objects is possible only because they are moving relative to objects outside them and contain other objects that are moving inside them. Matter and motion form the basic, inseparable, but nevertheless dialectical unity.

# CLASSICAL MECHANISM

Although the defeat of mechanism was engineered primarily by determinists, the spoils were reaped by indeterminists. In 1927, the great worldwide economic

---

[70] Engels, Frederick. *Dialectics of Nature*. Moscow: Progress, 1925, p. 70.

expansion, to which indeterminism was to owe its revival, had barely begun. The legitimate complaints of determinists were answered by a mishmash of new mathematical formulation and ancient superstition that met with ready public and governmental acceptance. The deterministic part of the testimony against classical mechanism was as follows:

## Deterministic Critique

As mentioned, in their battles with supernaturalism, mechanists went so far as to promise a **complete** description of the world. They attempted to answer the indeterminists' demands for proof by asserting that a final proof was possible. When Heisenberg showed that such **certainty** was not forthcoming, mechanism lost its title as the scientific worldview.

Because mechanism prevailed so long as the guiding philosophy of science, it accumulated an enormous indeterministic burden. Today's deterministic critique of mechanism amounts to a critique of old science and old beliefs that either were indeterministic to begin with or were, like the doctrine of absolute chance, the results of compromises with indeterminism. Except for MATERIALISM and, to a certain extent, INSEPARABILITY, the assumptions held by classical mechanism stand in opposition to the currently evolving scientific worldview. I will mention only a few of the major indeterministic aspects of classical mechanism.

First, mechanism has been historically and curiously associated with a static rather than a dynamic vision of the world.[71] This association developed in spite of the fact that the concept of motion sponsored initially by mechanism is as much a part of INSEPARABILITY as is the concept of matter. In practice, however, any model of the real, dynamic world would have to be static in comparison. Engrossed in their static models, mechanists tended to overemphasize things rather than processes.[72] Along with the static view, came the overemphasis on isolation and reversibility.[73] Instead of being dynamic and progressive, classical mechanism became static and conservative.

Second, classical mechanics, with which mechanism was tied, was restricted to mechanical terms such as "mass," "force," "velocity," and "acceleration," which allowed for only simple models. In keeping with their static, finite nature, and

---

[71] Rosnay, Joel de. *The Macroscope: A New World Scientific System*. New York: Harper & Row, 1979, p. 80.

[72] Cornforth, Maurice. *Materialism and the Dialectical Method*. New York: International, 1971, p. 41.

[73] Bryen, S.D. *The Application of Cybernetic Analysis to the Study of International Politics*. The Hague: Nijhoff, 1972, p. 6.

simplicity, these models did not allow for evolution.[74] In the last analysis, the motion they did permit was external to the models themselves. Whether the objects of concern were considered filled with the mysterious substance called "matter," or devoid of matter altogether, made little difference. The mathematical methods of classical mechanics implied that the whole was equal to the sum of its parts. Although such an approximation was sufficient for many engineering feats, it was inadequate for rapidly evolving systems—those clearly in motion within, as well as without. As a consequence, the rigidity of classical mechanism made it practically worthless in the social and biological sciences that began to grow explosively early in the 20th century.[75]

## Indeterministic Critique

The guts of the indeterministic complaint against mechanism always has been that there is more to the universe than matter in motion. The epithets rained down upon the mechanists who refused to leave room in their philosophy for the supernatural and its aficionados. Indeterminists made a space for themselves by denying the universality of the mechanical view. They only needed to point out that the analogy between machines and other systems was grossly inadequate. They could feel great indignation when the analogy was extended across the Cartesian line. For these folks it was usually enough to deride the "crass materialism which views reality as matter in motion"[76] If that didn't work, then the concepts of matter and motion could be cheapened and restricted in other ways.

Through the centuries, the slander of the concept of matter was so intense that its position as a pillar of science was continually in question. So it was, that fainthearted scientists of the late 19th century moved to disown matter and adopt pure motion instead. What began as a legitimate attack on its rigidity became an escape from mechanism via the newly developing concept of energy.[77] Indeterminists clamored for an outright substitution of energy for matter. Perhaps the greatest advocate of the switch was Wilhelm Ostwald, a physical chemist, who believed: "The ultimate goal of science is now presented as the task of establishing a world view consisting purely of energy concepts, without the use

---

[74] Conger, G.P. *New Views of Evolution*. New York: Macmillan, 1929, p. 193.

[75] Deutsch, K.W. *The Nerves of Government: Models of Political Communication and Control*. New York: Free Press, 1963, p. 30.

[76] Kwok, D.W.Y. *Scientism in Chinese Thought: 1900-1950*. New York: Biblo and Tannen, 1965, p. 66-67.

[77] Fleming, Donald. Introduction. Edited by Jacques Loeb, *The Mechanistic Conception of Life*. Cambridge, MA: Harvard University Press, 1964, p. xxvi.

of the concept of matter."[78] Staunch materialists vehemently criticized this antimechanistic use of the energy concept, but, as we will see in the section after next, this usage is still very much with us.

Doubts about the concept of matter, and in particular, its interconnection with the concept of motion are nothing new. It was to be expected that these doubts would become especially prevalent when simplistic, finite mechanism was yet to be replaced by a sophisticated, infinite form. Today, indeterministic scientists attack INSEPARABILITY, not so much by denying the concept of matter, or the concept of motion, but by denying the universality of the INSEPARABILITY of the two. The logical results of this denial once again are the dubious and inexplicable notions of matter without motion and motion without matter. Both are attempts to negate INSEPARABILITY and deserve illustration in some detail.

## ABSOLUTE ZERO: MATTER WITHOUT MOTION?

As an alternative to INSEPARABILITY, the concept of motionless matter has enjoyed a long and unfruitful career. At first the Earth itself was thought to be motionless, the center of a universe bounded by the fixed stars. But then, one after another, portions of the universe were shown to be in constant motion relative to each other. Since 1960, even the continents are officially recognized as being in motion with respect to each other. In every instance, the idea of matter in motion triumphed over the idea of matter without motion. On this point, Heraclitus remains victorious.

Why then, after so many defeats, hasn't the notion of absolute rest been put to rest? The reason lies in the fact that INSEPARABILITY is an assumption, and although, like CAUSALITY, it has passed numerous tests, it cannot be tested in the infinite number of cases that would be required to remove all doubt. Many of the motions of matter appear so slow that, to the naive realist, they are absolutely at rest. For instance, that great symbol of stability and permanence, the Rock of Gibraltar, at first glance appears to be completely motionless. To the sailors passing by, it is the ship that is in motion, not the Rock of Gibraltar. But we know that Gibraltar is part of a revolving and rotating planet and that it is composed of billions of atoms in which still more billions of electrons revolve around each nucleus a million billion times a second.

Until such detailed knowledge was obtained through scientific investigation, philosophers could assume, with the atomists, that, although the atom itself was

---

[78] Cassirer, *Determinism and Indeterminism*, p. 140.

always in motion, whatever was inside the atom was not. A similar view still exists concerning other aspects of the motions of atoms, especially those occurring near absolute zero.

If one is to challenge a fundamental assumption of science, it had better be done in an area where experimental evidence will be forever incomplete. The indeterministic interpretation of absolute zero is a good example of the modern-day rejection of INSEPARABILITY through the hypothesis of the existence of motionless matter. Such shenanigans have been going on throughout the 20th century where they achieved a new height in absurdity.

The theory of absolute zero concerns heat, usually defined as the vibratory motion of individual atoms of the elements. The measurement of temperature is the measurement of the rapidity and magnitude of this vibratory motion. If an atom had no vibratory motion, then it would exhibit no temperature. Ideally, we would call this temperature, or lack of temperature, absolute zero. I say "ideally" because in reality, no one has ever observed absolute zero. Claims have been made, but all these require an indeterministic interpretation of the Third Law of Thermodynamics.

Essentially, the Third Law "precludes the attainment of the absolute zero of temperature in a finite number of steps."[79] According to INSEPARABILITY, the absolute zero of temperature would imply the absence of matter as well as the absence of motion. Although absolute zero (about -273.15 degrees Celsius) may be a useful idealization, it is a mistake to speak of matter **at** absolute zero, because matter could not exist at that temperature. That is, it could not exist without being in motion. No experiment, which necessarily must be done in a finite context, could descend an infinite number of steps to finally achieve the "temperature" at which matter in motion does not exist.

As might be expected, the indeterministic interpretation of the Third Law generally confuses ideality with reality. It assumes that matter could actually exist **at** absolute zero, undergoing no motion, and having **perfect** order (zero entropy) and **perfect** crystallinity.[80] Even those who know that absolute zero cannot be obtained in a finite number of steps seem compelled to write in contradiction: "the absolute zero of temperature means no motion of the molecules."[81]

But it is indeterministic nonsense to dream of molecules at absolute zero. As mentioned, the atom itself consists of other objects in continuous motion. The electron path, which in any case is not perfectly circular, helps define the shape of

---

[79] Cohen, E.G.D. "Toward Absolute Zero." *American Scientist* 65 (1977): 752-58, p. 755.
[80] Cairns-Smith, A.G. *The Life Puzzle: On Crystals and Organisms and on the Possibility of a Crystal as an Ancestor.* Toronto: University of Toronto Press, 1971, p. 70.
[81] Cohen, "Toward Absolute Zero," p. 752.

the atom and varies from one revolution to another. The nucleus, too, responds to the position of the electron at any moment and therefore must wobble, as do the parts of a planetary system. A non-vibrating atom is therefore inconceivable.

It has been well established that entropy (amount of apparent disorder) **approaches** zero as the temperature of a substance **approaches** zero. In general, gases become liquids and liquids become solids as temperature decreases. Matter usually exhibits an increasingly well-ordered and generally more compact structure as it loses motion. For practical calculations in thermodynamics, the assumption that entropy is approximately zero near absolute zero is legitimate even though absolute zero is unreachable. Illegitimacy arises only when we incorrectly assume that the success of the approximation indicates the actual existence of the ideal. Even Lewis and Randall, the great authorities on thermodynamics, contributed to the confusion with this statement appearing in their classic text: "Every substance has a finite positive entropy, but at the absolute zero of temperature the entropy may become zero, and does so become in the case of perfect crystalline substances."[82, 83]

A logical consequence of this idealistic interpretation is the mistaken belief that the Third Law could "lay claim to perfect exactness."[84] Since Lewis and Randall sanctified it, others continue to speak of "crystals at absolute zero."[85] or perfection "near absolute zero."[86] We are to believe that "at the absolute zero of temperature…all thermal motion in a solid ceases and there can be no disorder due to lattice vibrations or other atomic motions."[87]

Such indiscreet statements leave the impression that absolute zero could actually be obtained, that motion then would cease, but that matter would somehow still exist—that matter without motion would be possible. We are seldom given much insight into the nature of such matter. Presumably, we must bring back the old conception of atoms as hard little balls that were once considered the ultimate

---

[82] Lewis, G.N., and Merle Randall. *Thermodynamics and the Free Energy of Chemical Substances*. New York: McGraw-Hill, 1923, p. 448.

[83] Lewis, G.N., and Merle Randall. *Thermodynamics*. (2 ed. of Thermodynamics and the free energy of chemical substances, revised by K. S. Pitzer and Leo Brewer). New York: McGraw-Hill, 1961, p. 130.

[84] Lewis and Randall, *Thermodynamics*. 1923, p. 461.

[85] Blum, H.F. *Time's Arrow and Evolution*. Princeton, NJ: Princeton University Press, 1968, p. 2.

[86] Whyte, L.L. *The Universe of Experience: A World View Beyond Science and Religion*. New York: Harper & Row, 1974, p. 53.

[87] Wood, B.J., and D.G. Fraser. *Elementary Thermodynamics for Geologists*. New York: Oxford University Press, 1976, p. 38.

constituents of matter. Perhaps that is what one commentator meant when he implied that "only at absolute zero would they be pure atoms."[88]

Such sophistry in evading the Fourth Assumption of Science led to a remarkable break through by W. G. Proctor.[89] Proctor suggested that matter really can exist at absolute zero because of "quantum mechanical considerations" further chasing the indeterminism into shadows where few dare to follow. After **assuming** that absolute zero had been achieved experimentally, Proctor even claimed to have discovered negative temperatures. By his own admission, however: "The experimental results run counter to a strong sentiment that temperature, like volume is something intrinsically positive."[90] Indeed, for those who believe that absolute zero can be achieved in the laboratory, the positivity of temperature must not be the only deterministic sentiment with which they have difficulty. Proctor's speculation on negative temperature actually applies a double whammy to INSEPARABILITY. Not only is matter seen here as motionless, but its motion rises ghostlike from its corpse. From matter without motion we are led to motion without matter.

## ENERGY: MOTION WITHOUT MATTER?

Ostwald's dream of replacing villainous matter with angelic motion is today being realized in the common mystification of energy as motion without matter. The situation has gotten so bad that even indeterminists have complained that "Physics has discarded matter, but has supplied no substitute."[91] As mentioned, "energy" is defined as the product of mass times the velocity of light squared—an attempt to account for both matter and motion at the same time. But the mere multiplication of a term for matter and a term for motion really does not guarantee their conceptual unification any more than the designation of matter and motion as separate terms guarantees their physical independence.

"Energy" and other matter-motion terms are easily misused because it is impossible to conceive of matter and motion as separate entities. We can observe matter *in* motion and the motion *of* matter. It is self deceptive to claim, as some do, that we can conceive of a thing as if it were motion and motion as if it were a thing. Legs are not motion and running is not matter. The inherent failing of a

---

[88] Hawkins, "The Thermodynamics of Purpose," p. 108.

[89] Proctor, W.G. "Negative Absolute Temperatures." *Scientific American* 239, no. 2 (1978). 90-99.

[90] Ibid., p. 99.

[91] Whyte, *The Universe of Experience*, p. 69.

matter-motion term is its tendency to subsume the connotations of matter at one time and those of motion at another, often unbeknownst to the user. Either way, the indeterminist can exploit the resulting confusion as a way of opposing INSEPARABILITY.

The fact that energy is matter-motion is forgotten in the all-too-common, but misleading view that matter is equivalent to energy.[92] This cannot be true because the term for matter (mass) in Einstein's equation never appears without the term for motion (velocity of light squared). It is forgotten again when energy and matter are viewed as mutually exclusive or as opposites.[93] This forces energy into the motion category, to which it is no better suited. These ambiguities are not resolved by lengthy exposition[94] and only demonstrate the difficulty of comprehending matter and motion as a singular phenomenon.

The concept of energy thus plays a unique role in science and philosophy. Determinists can see it as an admirable, if inevitably flawed attempt to unify conception and reality. Indeterminists can see its failure to do this as a proof that their necessarily divisible conception matches a divisible reality. Instead of supporting INSEPARABILITY, the idea that matter and motion are conceptually intertwined, the concept of energy has encouraged the opposite view, that matter or motion might be found physically independent of each other.

The idea that energy is motion without matter was natural for the positivistic approach to science that grew alongside the concept of energy. At the microscopic level of observation in particular, it is not uncommon to find evidence for motion without the corresponding evidence for its material carrier. At such a juncture one can **assume**, with the Fourth Assumption of Science, that the carrier exists, or one can **assume**, along with the positivists, that until one is found, a carrier does not exist. In the latter case, energy takes on a ghostly form unsuited to definition.

For example, physicists recognize the "flow" of heat as the mechanical transfer of motion from molecule-to-molecule within solid bodies. But when this same motion appears as infrared radiation, it is considered neither as matter, nor as the motion of matter, but as matter-motion, a mysterious, massless wave-particle capable of traveling through "empty space." In contrast, these same physicists hold a clear view of other types of motion. Sound, for instance, is not considered

---

[92] Daniels, Farrington, and R.A. Alberty. *Physical Chemistry*. 2 ed. New York: Wiley, 1961, p. 36.

[93] Nicolis, G., and I. Prigogine. *Self-Organization in Nonequilibrium Systems: From Dissipative Structures to Order through Fluctuations.* New York: Wiley-Interscience, 1977, p. 24.

[94] Hoffman, E.J. *The Concept of Energy*. Ann Arbor, MI: Ann Arbor Science Publishers, 1977.

matter or matter-motion. It is the motion of matter. Sound may undergo a similar transmission from a solid body to its gaseous surroundings without losing its character as motion. No one seriously dreams of the fantastic conversion of sound from motion into matter-motion, and yet this is exactly what is imagined in the case of heat motion. The result is the familiar conceptual muddle.

Duality can occur only between the concept of matter and the concept of motion, not between matter and motion itself. In thinking and writing about motion, we are often forced to use nouns instead of verbs although there are no material objects to which the concept applies. We may be forced to idealize matter and motion as if they were distinct entities, but we must never hypothesize their separation in reality. For example, it is nonsense to consider, as the Big Bang cosmogonists[95] do, that the universe was once devoid of matter, consisting only of radiation.[96]

The term "energy" is so frequently confused by its indeterministic interpreters that one may question its usefulness within a philosophical system guided by INSEPARABILITY. I suggest that the term "energy" be avoided whenever the more specific terms "matter" or "motion" could be used instead. Certainly, "energy" should never be opposed to matter or motion. Unfortunately, neither the term "matter" nor the term "motion" is completely unambiguous either. As explained later, each must be defined in terms of the other. As a noun describing action, the word "motion" unavoidably gives a nominative connotation to activity. Motion ends up being thought of as an entity. Instead of being what matter **does**, it tends to become what matter **is**. These problems are always with us and are best solved through practice in conceptualizing matter and motion.

## CONCEPTUALIZING MATTER AND MOTION

Hegel's insistence on the physical INSEPARABILITY of matter and motion remains valid today. In reality, matter and motion are one; only in ideality could they become distinct. But as Rosnay pointed out: "intelligence can understand movements or flows only as a succession of juxtaposed still positions."[97] Our sensations come in discontinuous pulses. Vision, for example, is a series of photon impacts that impart information about the shapes and qualities of the material

---

[95] Cosmogonists are cosmologists who assume that the universe had a beginning.

[96] Gamow, George. "The Evolutionary Universe." In *Cosmology + 1: Readings from Scientific American*, edited by Owen Gingerich, 12-19. San Francisco: Freeman, 1956 [1977], p. 18.

[97] Rosnay, *The Macroscope*, p. 161.

structures before us. The sensation of motion develops in the way in which thousands of still frames in a filmstrip produce a motion picture. Thus we may see matter, but we can only infer motion. Motion cannot be sensed, for it is not a thing. Only things can be sensed.

This is not to say that matter is real and that motion is unreal. Our idea of matter and motion stems from real matter and its real motion. But as noted before, the idea of existence applies only to matter. Only things exist, events do not. The Declaration of Independence exists, but the signing of it does not. The signing of the Declaration of Independence never "existed," it "occurred." The signing was the motion of matter, not matter itself.

If the first step in science is to distinguish one thing from another, then the second step is to distinguish the thing from what it does. A philosophy based on INSEPARABILITY seeks conceptual clarity by continually making the distinction between matter and motion. It explores the different types of motion as it explores the different types of matter. While it reduces reality to two abstract categories, it nevertheless insists that matter and motion form a basic unity that is concretely and infinitely varied in the types of phenomena it displays.

It always has been difficult to maintain the conceptual identities of matter and of motion while maintaining their physical INSEPARABILITY. It has become especially so with the widespread use of terminology that attempts to combine the two concepts but fails without warning. Fortunately, the basic structure of language can rescue us. In all languages, words refer to things or to what things do. Subjects naturally require predicates; predicates naturally require subjects. By insisting on clear distinctions between matter and what matter does, we can edify ourselves as we renew the language. When matter is being discussed, we can demand to know how it is moving; when motion is being discussed, we can demand to know what it is that is moving.

These questions, no longer pressed by physics, must be asked to produce clear thinking. We must rediscover which words refer to matter and which refer to motion. It is easy to classify "legs" and "running," but there are many words, including some general terms, that have carried the burden of indeterministic obscurantism. I will define and discuss a few of them to initiate the discipline essential for applying the assumption of INSEPARABILITY. The basic strategy is this: separate matter and motion into distinct **conceptual** categories, and after having done so, demand a marriage that will not allow one of them to appear without the other. Finding matter, we will ask: "Where is motion?" Finding motion, we will ask: "Where is matter?"

Table 4-1. Some common alternate terms for the categories "matter" and "motion."

| Matter | Motion |
|---|---|
| Thing | Event |
| Structure | Function |
| Mass | Velocity |
| Space | Time |

## Thing-Event

All things are matter and all events are motion. In a strict sense, we should not speak of "things occurring," because only events can occur. Things simply exist. An "event" is the convergence or divergence of at least two things.

## Structure-Function

Structure and function are particularly common terms in biology and sociology. Even though structure and function are inseparable,[98] entire disciplines have managed to emphasize one or the other as though they occurred independently. Sociology and anthropology, for example, have two philosophically opposed schools: the "structuralists" and the "functionalists." Aside from the delicate connotations and historical nuances developed in the debates between these two groups, it must not be forgotten that, if the starting point is a bogus terminological distinction, then much of the debate and much of the science flowing from it also will be suspect. A discussion between a pure "matterist" (one who believed in matter without motion) and a pure "motionist" (one who believed in motion without matter) would have little more than humor to recommend it. That the word "structure" can be found so remote from the word "function" betrays a rejection of INSEPARABILITY and an indication that the language of indeterminism is being spoken.

## Mass-Velocity

As a student in elementary physics I was shocked to find that something as basic as "mass" had no *a priori* foundation. As it turns out, the definition of mass

---

[98] Nagel, Ernest. *The Structure of Science*. New York: Harcourt Brace and World, 1961, p. 425.

is far from a trivial problem.[99] We must consider mass as matter and velocity as motion. The matter-motion term for mass-velocity is momentum, p. The equation being:

$$p = mv \qquad (4\text{-}1)$$

Where:
m = mass, g
v = velocity, cm/sec

Because momentum is neither mass nor velocity, neither matter nor motion, it is subject to the same conceptual confusion as is the term "energy." Objects of great mass and low velocity relative to the observer take on the connotations of matter, while objects of small mass and high velocity relative to the observer take on the connotations of motion. In equation 4-1, mass is defined in terms of momentum and velocity, while velocity is defined in terms of momentum and mass—a circular argument made necessary by the reality of their INSEPARABILITY. Just as there is no *a priori* in philosophy, there can be no *a priori* in nature, no absolute mass to which all others naturally relate. There can be no mass independent of velocity, just as there can be no matter independent of motion.

The isolation implied when velocity is stated without a referent simply does not exist. An "object" surrounded by "empty space" would have no mass just as it would have no velocity. Mass, like velocity, is dependent on the existence and motion of other things. While equation 4-1 yields a velocity of zero for an object at "rest" relative to the Earth, its velocity certainly is not zero with respect to other objects in the universe. By keeping the necessity of a referent in mind, equation 4-1 becomes just another way of stating INSEPARABILITY: "objects" without motion relative to other objects in the universe would have no momentum, and by implication, no mass either. They simply would not exist.

## Space-Time

Much of modern physics and cosmology concerns the discussion of space and time. The resulting confusion reflects one of the great philosophical struggles of the 20th century. We will need some carefully honed definitions to make any sense of it. Because the ones I choose here—space is matter, time is motion—will seem counterintuitive to many of you, let me explain my reasoning.

---

[99] Ibid., p. 192-196.

Under MATERIALISM, we assumed that the universe consists of matter. As mentioned, "matter" is defined as an abstraction for "all things." Every thing we are familiar with, and presumably those we are not, takes up space. Everything in existence has dimensionality, and therefore volume. If the universe is material, then the volume or space it occupies must be regarded as matter. The term space, used is this way, becomes another abstraction for "all things," i.e., matter. For the determinist, space is something; for the indeterminist, space is nothing.

The ancient, commonsense idea of opposing space to matter as though space was the absence of matter is no longer as useful as it once was.[100] For one thing, pure space has been found nowhere. Experimentally, the closest we have ever come to empty space is a partial vacuum. For another, matter takes up space and would be inconceivable without space. In addition, matter is generally found to consist mostly of what we generally consider to be space. For example, the mass of an atom is almost all located within the nucleus. Its volume is mostly "empty space." However, when we investigate this so-called "empty space" we invariably find that it is not empty at all. There is always something there.

Only in ideality could matter and space be considered opposites. Perfectly solid ideal matter is the opposite of perfectly empty ideal space. In reality, neither solid matter nor empty space has been found, disappointing naïve realists whenever they discover this fact. All things and their surroundings evince varying degrees of solidity and emptiness. A completely empty universe thus becomes no more conceivable than a completely solid one. Thus the definition of space as the absence of matter is as obsolete as the definition of matter as hard and impenetrable.

Time is motion. In the specific, time is the motion of one thing relative to another thing. In all cases, we measure time by measuring the motion of things. Universal time is the motion of each thing relative to all other things. There can be no separate existence for time just as there can be no separate existence for motion. The universe is already filled with matter. There is no "room" for motion in the universe and, as we will see much later, it is unnecessary to create another dimension for it. There never was a "time" when the universe did not exist. Speculations about going "back in time" are mere science fiction amusements. There is no such place.

As with other matter-motion terms, the concept of "spacetime" is an attempt at visualizing an inseparable reality. It cannot be done, of course. We can only visualize a portion of the universe and then visualize the motion of it in relation to another portion of the universe. One way out of this is to separate the words as we separate the concepts of space and time in our minds. To remind ourselves of

---

[100] Cassirer, *Determinism and Indeterminism*, p. 125.

this necessary conceptual separation, let us henceforth place a hyphen between "space" and "time."

One might have thought that the development of matter-motion terms in modern physics would have laid **separability** to rest. Not true. The notion that "spacetime" can be modeled in four dimensions assumes that "spacetime" can be treated as matter. Dimensionality is a property of matter, but cannot be a property of motion, because motion does not *exist*, it *occurs*. Like other terms for matter-motion, "spacetime" tends to take on the connotations of either matter or motion at the whim of the user. When an attempt is made to build a model of "space-time," the model, being material, tends to give a material connotation to "space-time". When an attempt is made to conceive of "space-time" as motion, the concept of the universe tends to become dematerialized. The notion that the universe might be "curved" thus is not far removed from the notion that it does not exist at all.

## INSEPARABILITY AND CLEAR THINKING

Let me restate. The dialectical nature of the world stems from its character as matter in motion. Its unity consists in the INSEPARABILITY of this matter and its motion. Although matter and motion are not physically separable, it is impossible for the mind to conceive of matter and motion as a singular phenomenon. Although we may invent terms for conceiving of matter-motion as a unity, they inevitably fail, taking on the connotations of either matter or motion, not both at once. Clear thinking requires us to be cognizant of INSEPARABILITY. Consequently, we must guard against four types of errors of logic that violate the assumption of INSEPARABILITY:

1. That matter could occur without motion.
2. That motion could occur without matter.
3. That matter **is** motion.
4. That motion **is** matter.

Only by avoiding these indeterministic errors can we achieve a description of the universe that includes both subject and predicate, and is therefore both meaningful and scientific.

# CHAPTER 5

# THE FIFTH ASSUMPTION OF SCIENCE: CONSERVATION

*Matter and the motion of matter neither can be created nor destroyed.*

The sentiment underlying CONSERVATION is ancient. The Greek philosopher, Anaximander, asserted that matter is eternal and indestructible. The Roman philosopher, Lucretius, believed that "Never can nothing become something, nor something nothing." Although we have since discovered that the individual units of matter are not as permanent as these early atomists thought, the deterministic notion that the external world has an overall permanence still persists. David Bohm clearly expressed the modern view:

> In nature nothing remains constant. Everything is in a perpetual state of transformation, motion, and change. However, we discover that nothing simply surges up out of nothing without having antecedents that existed before. Likewise, nothing ever disappears without a trace, in the sense that it gives rise to absolutely nothing existing at later times. This general characteristic of the world can be expressed in terms of a principle which summarizes an enormous domain of different kinds of experience and which has never yet been contradicted in any observation or experiment, scientific or otherwise; namely everything comes from other things and gives rise to other things.[101]

The alternative to CONSERVATION is **creation**, the assumption that material entities can come into being without material antecedents. In its greatest generality **creation** supposes that a rational but immaterial being existed by itself for an eternity before it resolved to create the universe out of nothing. The latest version

---
[101] Bohm, *Causality and Chance in Modern Physics*, p. 1.

among those who call themselves creationists is the argument for "intelligent design."[102]

In day-to-day application, of course, only tiny portions of the universe are considered liable to this supernatural way of doing things. The belief in miracles, for instance, is an attempt to make the idea of **creation** a living reality rather than a remote suspicion. Even where the belief in gods and miracles is in decline, more sophisticated forms of creationism arise to take its place.[103] For example, in the United States up to 90% of the citizens[104] and over half of the scientists believe in extrasensory perception (ESP)[105] even though there isn't a shred of scientific evidence for it.[106]

All manner of occult beliefs remain ever popular even though they have repeatedly failed the simplest of scientific tests. Psychics,[107] astrologers,[108] and psychic healers[109] actually have taken a good drubbing in the popular press, but the ignorance keeps on coming with the birth of each child. For the most part, such stories are overlooked on the way to the astrological charts that serve as daily fare. If anything, the prevalence of occult belief, and especially its new-age adoption in the form of eastern mysticism within the very heart of modern physics,[110] shows **creation** to be a viable alternative to CONSERVATION.

# FROM ATOMISM TO EVOLUTION

The idea of conservation evolved out of the primitive idea of matter. The atomists assumed that matter consists of hard little balls composed of an indivisible material. It was these bits of matter, then, that were conserved. They had a stability

---

[102] Scott, E.C. "Not (Just) in Kansas Anymore." *Science* 288, no. 5467 (2000): 813-15.

[103] Abell, G.O., and Barry Singer. *Science and the Paranormal: Probing the Existence of the Supernatural.* New York: Scribner, 1981; Singer, Barry, and V.A. Benassi. "Occult Beliefs." *American Scientist* 69 (1981): 49-55.

[104] Singer and Benassi, "Occult Beliefs," p. 49.

[105] Ibid., p. 54.

[106] Hansel, C.E.M. *ESP: A Scientific Evaluation.* New York: Scribner, 1966.

[107] Randi, James. "Geller a Fake, Says Ex-Manager." *New Scientist* 78 (1978): 11.

[108] Fraknoi, Andrew. "Astrology Put to the Test." *San Francisco Chronicle (Sunday Examiner World)*, December 9 1979; ———. "Further Tests of Astrology." *San Francisco Chronicle*, April 14 1981.

[109] Saltus, Richard. "Psychic Snooper Gets the Goods on 'Miracle' Workers." *San Francisco Chronicle (and Examiner)*, August 24 1980.

[110] Capra, Fritjof. *The Tao of Physics: An Exploration of the Parallels between Modern Physics and Eastern Mysticism.* New York: Bantam Books, 1975.

or permanence transcending that of the objects of which they were parts. In the extreme, these atoms were considered indestructible and eternal. As recently at the 18th century, this simple belief in the conservation of matter successfully guided great scientific achievements.

For example, Antoine Lavoisier, often called the father of modern chemistry, noted that certain materials gained weight when ignited. He assumed that this added weight was a result of added matter that existed prior to combustion. While a creationist might have insisted that the increase in weight was indeed a miracle—that something had been created out of nothing—Lavoisier looked to the atmosphere for the missing constituent: oxygen. Lavoisier then showed that certain elements always combined with certain other elements in fixed ratios. For instance, 12 grams of carbon combined with 32 grams of oxygen to form 44 grams of carbon dioxide. For a long time, conservation, conceived exclusively as the conservation of matter, achieved success after success.

At first, when it was noticed that every chemical reaction either emitted or absorbed heat, it was logical to assume that the heat was just another form of matter: the caloric fluid. The caloric theory was a simple extension of the assumption that matter is always conserved.

What was missing from the law of conservation, of course, was the same concept that was missing from materialism prior to INSEPARABILITY: motion. Here again we see the lag in the development of the idea of motion. Because motion was not a thing, its conservation could not be viewed in the same, simple way that an atomist could view the conservation of matter. The idea of its conservation could not arise until Newtonian dynamics began to show how.

Once heat was viewed as motion rather than matter, there was no denying the inclusion of all other forms of motion within the framework of CONSERVATION. Today, conservation is usually stated as the First Law of Thermodynamics, otherwise known as the law of the conservation of energy: "Energy may be transformed from one form to another, but it cannot be created or destroyed, and the total energy of an isolated system is constant."[111] Although energy is a matter-motion term, one suspects that it is motion instead of matter that now has gotten the upper hand in this modern statement of conservation. That is why I prefer to state CONSERVATION in a way that is clearly compatible with INSEPARABILITY: **Matter and the motion of matter neither can be created nor destroyed**. The following reaction illustrates the practical use of this conception of CONSERVATION. Methane gas burns in the presence of oxygen with the production of carbon dioxide and water and the emission of heat:

---

[111] Daniels and Alberty, *Physical Chemistry*, p. 36.

$$CH_4 + 2O_2 + kcal \longrightarrow CO_2 + 2H_2O + 212 \text{ kcal} \qquad (5\text{-}1)$$

methane oxygen activating   carbon   water   heat
                         motion    dioxide              motion

Note that the number of carbon [C], oxygen [O], and hydrogen [H] atoms before and after the reaction is the same. All that happens during the reaction is the conversion of one form of matter in motion into another form of matter in motion. Hydrogen diverges from its combination with carbon in methane to converge on and combine with oxygen as water. Some of the oxygen also converges on carbon to form carbon dioxide. Some of the rapid vibrations of methane and oxygen molecules have been converted into the slower vibrations of carbon dioxide and water and the faster motions of whatever matter exists in the surroundings.

From equation 5-1 it should be clear that what is being conserved is both matter and motion. One type of matter in motion is being changed into another type of matter in motion; one type of the motion of matter is being changed into another type of the motion of matter. The conventional use of the matter-motion term "energy" to describe conservation might be admirable, if it did not end up being philosophically misleading.

## CHALLENGES TO CONSERVATION

As with the other assumptions of science, CONSERVATION has met many challenges from the opposing viewpoint. For obvious reasons, it is seldom admitted that CONSERVATION implies that the universe is eternal. Still, whoever uses this assumption invariably conflicts at some point with those who hypothesize a "first cause" for the universe. The **creation** argument is indeterministic, not only because the so-called causative agent that it assumes is immaterial, but also because it assumes that something can be created out of nothing. This agent is therefore empty, vacuous, without integrity and without the possibility of existence. Creationists seldom agree on where in the causal chain of natural events this supernatural activity should be inserted. Some insist that **creation** is occurring at this very moment, while others insist it occurred 13.7 billion years ago. Except for its temporary suspension in conventional cosmology, however, CONSERVATION has met the challenge and succeeded in pushing **creation** from serious scientific discourse. This is especially true now that the concept of motion has survived the 19$^{th}$ century firestorm engendered by its inclusion in the law of conservation.

## Geology

Some of the earliest battles over CONSERVATION took place in the study of geology, where field evidence indicated that the Earth was much older than Biblical accounts of **creation** would have it.[112] CONSERVATION entered the debate in the form of "uniformitarianism," the assumption that geological processes occurred in the past at the same rate as at present. To squeeze their interpretations into the six thousand years then allowed by church doctrine, sincere geologists invented an opposing assumption: "catastrophism" the view that geological processes were much more rapid in the past than at present. With catastrophism, sedimentary sections hundreds of meters thick still could be attributed to the Biblical flood.

Although it is true that some geological processes are indeed extremely rapid, this did not explain the fossil evidence. Fossils collected from the upper portions of the sedimentary sections were more like those of today than the ones from the lower portions. Many species had changed noticeably during the deposition period. Forty days and forty nights didn't come close. Some creationists figured that these changes were brought about by individual miracle and could still fit the time frame of the deluge. Still others saw the progression of the fossils as the result of acts of the devil meant only to confuse the weak of faith. The new version of CONSERVATION brought the conflict closer and closer to biology, preparing the way for Darwin. In 1841 Hugh Miller pointed out the terrible choice that confronted would-be scientists: "There is no progression. If fish rose into reptiles, it must have been by sudden transformation.... There is no getting rid of miracle in the case—there is no alternative between creation and metamorphoses. The infidel substitutes progression for Deity; Geology robs him of his god."[113]

## Biology

In biology, uniformitarianism found its parallel in the doctrine of evolution. Throughout science the belief in the conservation of **things** evolved into the belief in the conservation of **processes**. The full implication was that nothing was immutable; absolutely everything was in motion, evolving into other things. Whereas some past notions of conservation might have been supportive of **creation**, this no longer remained the case.

---

[112] Gillispie, C.C. *Genesis and Geology: A Study in the Relations of Scientific Thought, Natural Theology, and Social Opinion in Great Britain, 1790-1850.* New York: Harper Torchbooks, 1951.

[113] Miller, Hugh. *The Old Red Sandstone.* 7th ed. Boston: Gould and Lincoln, 1841, p. 41.

The revolution was marked by the publication of Charles Darwin's classic in 1859 that unleashed the greatest of all battles between science and religion.[114] Won overtly by religion and covertly by science, the evolution-creation struggle continues to this day. As recently as the Scopes trial of 1927, creationists were successful at preventing the teaching of evolution in science courses, at least at the elementary levels. Of course for scientific research to advance at all, some form of conservation doctrine was an absolute necessity. Indeterministic control generally has been weak at the college level, where the teaching of evolution has frequently provided a revelation for those who were only warned of its vices while in high school.[115] Today the concept of evolution is occasionally taught in the lower grades as part of the science curriculum, where it is free to confront creationists at an early age. In the US, evolutionary ideas are no longer restricted to the college educated, except in the most backward of communities.

Skirmishes between evolutionists and creationists have been especially hot topics in the U.S. during the last two decades,[116] reaffirming the undying opposition between the assumptions of CONSERVATION and **creation**. Because scientific advancement without the concept of evolution is no longer possible, these debates now serve mostly as opportunities for educating the public.

## Cosmology

In cosmology, theories of the origin of the solar system went through a similar evolution.[117] Newton, in keeping with his atomistic views, believed that the immutable objects of the solar system were set into motion by the "force" of the creator. As late as 1781, even the irreligious French naturalist, Comte de Buffon, supported **creation** as the logical origin of the solar system. In the 19th century, Laplace, following Immanuel Kant, proposed the nebular hypothesis for its origin. At first this threatened a science-religion conflict almost as vehement as the one in biology,[118] but the idea that the sun and planets accreted from a nebular cloud

---

[114] Eiseley, Loren. *Darwin's Century: Evolution and the Men Who Discovered It*. Garden City, NY: Doubleday Anchor, 1958.

[115] Field, M.D. "Poll Shows State Favors Teaching of Evolution." *San Francisco Chronicle*, May 15 1981, p. 9.

[116] Kitcher, Philip. *Abusing Science: The Case against Creationism*. Cambridge, MA: MIT Press, 1982; Scott, "Not (Just) in Kansas Anymore"; Pigliucci, Massimo. *Denying Evolution: Creationism, Scientism, and the Nature of Science*. Sunderland, MA: Sinauer Associates, Inc., 2002.

[117] Numbers, R.L. *Creation by Natural Law: Laplace's Nebular Hypothesis in American Thought*. Seattle: University of Washington Press, 1977.

[118] Ibid., p. 88.

readily invoked images like those in Biblical accounts. As always, natural order also could be viewed as the handiwork of an omniscient creator. Within every conflict lies the possibility of peaceful settlement: "The potential threat to scientific progress posed by the insistence on agreement between science and the Bible failed to materialize largely because pious and ingenious men repeatedly succeeded in devising new ways to reconcile the two revelations."[119] In this endeavor astrophysics currently leads the way,[120] but not without resistance.[121]

The lesson is clear: no matter what deterministic scheme is proposed, it can always be diluted to make its implications more palatable to the indeterminist.

The assumption of **creation** may have been pushed out of the solar system, and studies of galactic evolution[122] may have pushed it still further, but it has not been removed from scientific thought altogether. We find that scientifically credible theories of the universe still may be classified either as: "cosmologies," which, in keeping with CONSERVATION, assume that the universe had no origin, or as "cosmogonies," which, in keeping with creation, assume that it did. Despite the special pleas that can be made for it,[123] today's most popular theory, the standard Big Bang, is of the second type. As you will note throughout this discussion, the Big Bang Theory is a blatant violation of many other scientific assumptions in addition to CONSERVATION.

# FROM THE STATIC TO THE DYNAMIC

The struggle to include motion in the law of conservation mirrored the development of INSEPARABILITY. Just as MATERIALISM had to be supplemented with CAUSALITY to advance from materialism to determinism, the concept of matter had to be united with the concept of motion in the assumption of CONSERVATION. The static materialistic worldview thereby gave way to a dynamic materialistic worldview. As an indeterministic alternative, **creation** remains as useless as it ever was for guiding scientific investigation. The hypothesis of an immaterial "first cause" for any particular phenomenon really is not a beginning point, but an ending point.

---

[119] Ibid., p. 88.

[120] Ross, Hugh. *The Creator and the Cosmos: How the Latest Scientific Discoveries of the Century Reveal God.* 3rd ed. Colorado Springs, CO: NavPress, 2001.

[121] Kurtz, Paul, ed. *Science and Religion: Are They Compatible?* Amherst, NY: Prometheus Books, 2003.

[122] Strom, S.E., and K.M. Strom. "The Evolution of Disk Galaxies." *Scientific American* 240, no. 4 (1979): 72-82.

[123] Cherfas, Jeremy. "Evolution: Survival of the Creationists." *New Scientist* 89 (1981): 128-29.

# CHAPTER 6

# THE SIXTH ASSUMPTION OF SCIENCE:
# COMPLEMENTARITY

*All bodies are subject to divergence and convergence from other bodies.*

With INSEPARABILITY, and again with CONSERVATION, the external world was viewed as matter in motion. COMPLEMENTARITY continues with this point of view, interpreting the Second Law of Thermodynamics (SLT) as a law of divergence and its so-far unrecognized complement as a law of convergence. Only by assuming COMPLEMENTARITY can we resolve the contradiction between CONSERVATION, that assumes that the universe is eternal, and the indeterministic interpretation of the SLT, that implies that it is not.

The alternative, **noncomplementarity**, assumes that matter or motion (by itself) can diverge from one part of the universe without converging on another part. It hypothesizes an ever-increasing disorder in the universe, but cannot adequately explain the increases in order we see all around us. This last difficulty, peculiar to today's system-oriented view, I call the "SLT-Order Paradox." Through its resolution the idea behind COMPLEMENTARITY will become clear.

## THE SLT-ORDER PARADOX

The Second Law of Thermodynamics (SLT) states that **the entropy or apparent disorder of an ideally isolated system can only increase**. In the strictest sense, the SLT says everything about increasing disorder, but nothing about increasing order. Yet as philosopher-physicist L. L. Whyte noted: "The fact which we cannot, it seems, deny is that over vast regions of space and immense periods of time the tendency toward disorder has not been powerful enough to arrest the formation of the great inorganic hierarchy and the myriad organic ones."[124]

Indeed, it appears that for every system in which order is decreasing, there is another in which order is increasing. The SLT predicts only destruction, while nature exhibits construction as well—the SLT-Order Paradox. The Second Law of Thermodynamics obviously tells only half of the story.

The other half of the story is still to be explained by a principle that complements the SLT. Many investigators[125] have recognized that the SLT by itself is inadequate for resolving the SLT-Order Paradox and for explaining the source of order. They obviously have not been completely satisfied with the conventional resolution of the paradox, which is generally stated like this: "whenever a semblance of order is created anywhere on Earth or in the universe, it is done at the expense of causing an even greater disorder in the surrounding environment."[126]

This implies that a finite, isolated universe would run down like a clock. In the popularized view, the universe is descending deeper and deeper into chaos as the order in the surroundings of every system is exhausted.[127] This prospect causes philosophical unease among scientists because it implies an initial creation as well as an eventual "heat death" of the universe. We require some principle that would both complement the SLT and avoid this predicted violation of CONSERVATION, the assumption that matter and the motion of matter neither can be created nor destroyed. There are no scientifically verified exceptions to either the First or the Second Law of Thermodynamics. And yet, there is still no adequate explanation for the apparent production of order from disorder.

Clearly, to resolve the SLT-Order Paradox we must have a radical departure from the present theoretical approach to the problem rather than a change in experimental technique or calculation. If the ending predicted by the current interpretation of the SLT is unacceptable, then there must be something wrong with its initial assumptions.

At this point it may be helpful to explain briefly what scientists mean when they speak of an "isolated system," "controlling an experiment," or "closing the

---

[124] Whyte, *The Universe of Experience*, p. 47.

[125] Schroedinger, Erwin. *What Is Life? The Physical Aspect of the Living Cell and Mind and Matter*. New York: Cambridge University Press, 1967; Whyte, *The Universe of Experience;* Makridakis, Spyros. "The Second Law of Systems." *International Journal of General Systems* 4 (1977): 1-12; Nicolis and Prigogine, *Self-Organization in Nonequilibrium Systems;* Prigogine, Ilya. "Time, Structure, and Fluctuations." *Science* 201 (1978): 777-85.

[126] Rifkin, Jeremy. *Entropy: A New World View*. New York: Bantam, 1980, p. 6.

[127] Ibid.

70 • The Ten Assumptions of Science

doors" on a portion of the universe. In conventional scientific terminology, the closest thing to "a portion of the universe" is called a "system," any object or group of objects that the investigator wishes to consider and to delineate in some way.[128] Ideally, systems can be of three types: isolated, closed, or open. "Isolated" systems exchange neither matter nor motion with the environment. "Closed" systems exchange motion but not matter. "Open" systems exchange both matter and motion. These definitions are idealizations developed from the study of relatively isolated and relatively closed systems. In reality, all systems are open systems; truly isolated or closed systems cannot exist.

Although competent scientists no longer believe that any real system could be ideally isolated, few of them seem prepared for the next step: the concept of ideal nonisolation. COMPLEMENTARITY assumes that real systems exist between the extremes of ideal isolation and ideal nonisolation. Whereas a high degree of isolation implies minimum contact between the system and its environment, a high degree of nonisolation implies maximum contact between the system and its environment.

Science has traditionally emphasized one end of this continuum: the system, isolation, increasing disorder, and the SLT. We need to emphasize the other end too: the environment, nonisolation, increasing order, and the complement of the SLT. The resolution of the SLT-Order Paradox awaits a balanced consideration of both the system and its environment. If this analysis is correct, then traditional, system-oriented attempts at resolution are bound to fail, as a few notable examples will demonstrate.

## System-Oriented Rationalizations of the Paradox

Each system-oriented attempt to resolve the paradox fails to the degree that it favors the system over the environment. Note in each of the examples, that whether the proposal involves unabashed vitalism,[129] the "geometry of space-time",[130] outright contradiction,[131] or sophisticated neovitalism,[132] the key to the production of order, the environment, is slighted.

---

[128] Weinberg, *An Introduction to General Systems Thinking*, p. 63.
[129] Schroedinger, *What Is Life?*
[130] Whyte, *The Universe of Experience*.
[131] Makridakis, "The Second Law of Systems."
[132] Nicolis and Prigogine, *Self-Organization in Nonequilibrium Systems*; Prigogine, "Time, Structure, and Fluctuations."

## Schroedinger (1967)

In addition to his work on wave equations in quantum mechanics, Erwin Schroedinger is known for his popularization of the concept of "negative entropy" or negentropy as a resolution of the SLT-Order Paradox.[133] In itself, the idea of an ordering process that functions as the dialectical opposite of the disordering process is excellent. The term "negentropy" is likewise excellent. What must be objected to is the biased way that Schroedinger described the negentropic process.

Negentropy was seen as a "fight" in which organisms, by themselves, overcame the havoc of the phenomena described by the SLT. The argument essentially followed the philosophical tradition of vitalism: neither matter nor the motion of matter was considered the initiator of the negentropic struggle. The mysterious source of order was internally derived, and was peculiar to living beings. Not only did Schroedinger overemphasize the system itself as a source of order, but he left the SLT-Order Paradox unresolved, at least wherever life was not evident.

## Whyte (1974)

A slightly improved attempt to resolve the apparent contradiction between the SLT and the tendency toward increases in order was made by L. L. Whyte.[134] Unlike Schroedinger, Whyte was careful to include the inorganic as well as the organic realm in his suggestion. Like Schroedinger, Whyte recognized the need for COMPLEMENTARITY when he wrote of the "two great, and **apparently opposed**, general tendencies."[135] Unlike Schroedinger, Whyte did not overtly confine his search for the source of increasing order to the system itself. Instead, he tried to avoid consideration of system-environment interactions through an approach that was more in tune with modern physics than with systems analysis.

Whyte's suggestion is puzzling. The first of the opposed tendencies involved matter and was "TOWARD DYNAMICAL DISORDER called **Entropic**."[136] The second involved geometry and was "TOWARD SPATIAL ORDER called Morphic."[137] Just how matter and geometry can be seen as independent features of the universe was not explained. As far as I can tell, the "Morphic" tendency seems to have much in common with "curved space" in the general theory of relativity. It

---

[133] Schroedinger, *What Is Life?*
[134] Whyte, *The Universe of Experience.*
[135] Ibid., p. 42.
[136] Ibid., p. 42.
[137] Ibid., p. 42.

explains the tendency toward order in one of the ways Einstein explained gravitation. The "geometry of space" purportedly supplies the orderly, passive fabric upon which the SLT operates. Whyte's answer to the SLT-Order Paradox requires the inscrutable interaction of matter with the supposed 4-dimensional geometry of "spacetime" rather than the interaction of matter with matter.

## Makridakis (1977)

Spyros Makridakis, a management scientist specializing in General Systems Theory, took his shot at the paradox by rightly claiming that the exact opposite of the SLT was as natural as the SLT itself.[138] But then he proceeded to get it backwards. According to Makridakis, the Second Law of Systems resolved the SLT-Order Paradox: "things tend to become more orderly if they are left to themselves." The phrase: "left to themselves" normally means that there is no outside interference. Of course, any system not subject to any outside interference whatsoever is an ideally isolated system. Rather than being a complement to the SLT, this suggestion was merely a contradiction of it. The opposition between the SLT and its complement cannot be derived by viewing systems in their isolation, but in their nonisolation. With respect to the SLT, Makridakis carried systems philosophy to its logical conclusion. The only thing that would save the Second Law of Systems would be to change it to read: "things become more orderly if they are **not** left to themselves."

## Prigogine (1978)

Perhaps the most celebrated approach to the SLT-Order Paradox within the discipline of thermodynamics was developed by Nobelist Ilya Prigogine.[139] While Schroedinger, and again, Makridakis, unabashedly treated systems in the customary way (as isolated entities providing their own source of order), Prigogine took some of the early steps toward viewing the environment rather than the system as a source of order.

Prigogine's challenge to classical thermodynamics suitably stressed that complex structures can exist only through continuous interaction with their surroundings. Without this interaction, structures tend to "dissipate," that is, they lose matter or motion as per the SLT. Following Onsager,[140] Prigogine developed

---

[138] Makridakis, "The Second Law of Systems."

[139] Prigogine, "Time, Structure, and Fluctuations"; Nicolis and Prigogine, *Self-Organization in Nonequilibrium Systems*; see also: Procaccia, Itamar, and John Ross. "The 1977 Nobel Prize in Chemistry." *Science* 198 (1977): 716-17.

[140] Onsager, Lars. "Reciprocal Relations in Irreversible Processes. I." *Physical Review* 37 (1931): 405-26.

the principle of minimum entropy production. His most important conclusion: there had to be a relationship between the production of order and the prevention of disorder.

Unfortunately, due to the constraints of the paradigm—systems philosophy—under which Prigogine and almost all modern scientists work, this did not lead directly to a singular principle that could be considered fully complementary to the SLT. Prigogine eventually was led to suggest some silly producers of order: fluctuations, distance from equilibrium, and nonlinearity that were not explicitly system-environment interactions. In the end, they had to be considered subsystem interactions.

Despite all his mathematical acrobatics, Prigogine's mechanisms could not be considered net producers of order for the system as a whole in the same way that phenomena described by the SLT produce disorder for the system as a whole. Thus, fluctuations produced as a result of interactions between the system and its environment eventually ended up being attributed to the system itself. Similarly, equilibrium and nonlinearity were said to occur **in** the system rather than **between** the system and its environment. There was always a residual bias in favor of the system over the environment.

Like Schroedinger, Whyte, and Makridakis, Prigogine offered reasons for the production of order in opposition to the SLT from a system-oriented point of view. Following tradition, he ultimately focused on the system—the forte of the SLT—to the neglect of the environment. He insisted that the production of order is a "self-organizing" process—a sort of neovitalism that, although not restricted to living systems, ultimately neglects environmental factors as producers of order. In my view, the ideal of nonisolation is equally as important as the ideal of isolation. Because such belief is, by definition, foreign to systems philosophy it cannot produce a complement to the Second Law of Thermodynamics.

## Resolution of the Paradox

Systems philosophy was adequate for developing the SLT, a law about ideal isolation. An environmentally focused viewpoint would permit the development of a complementary principle, a law about ideal nonisolation. The unification of these two one-sided viewpoints must consider both systems and their environments as equally important. **The SLT-Order Paradox can be resolved only through a balanced system-environment approach that describes the reality existing between ideal isolation and ideal nonisolation.**

Actually, an early step in this direction had been taken long ago by classical mechanics. According to Newton's First Law of Motion: "Every body perseveres in its state of rest, or of uniform motion in a right line, unless it is compelled to change that state by forces impressed thereon.[141] Like those who later developed thermodynamics, Newton first assumed that his system was ideally isolated. Newtonian bodies traveled through empty space or the stationary ether under their own inertia. Then, on second thought, he discarded the notion of ideal isolation and completed his First Law of Motion. Classical thermodynamics managed the first thought but not the second.

In devising the SLT, the originators of thermodynamics also assumed the system to be ideally isolated—it was necessary to be temporarily myopic. But if we should now reject this system myopia as Newton attempted to do, we would have a pertinent question to ask: "If matter or the motion of matter has diverged spontaneously from such an "isolated system" where has it gone?" The obvious answer is that it has moved toward other matter in the universe. If the universe was infinite, there would be no perfectly isolated systems; all matter everywhere would converge on and diverge from matter everywhere else.

If the above statement is true, then Newton's First Law of Motion must be modified to reflect this balance—the word "unless" must be replaced by the word "until." This small adjustment completes the train of thought that Newton only began and classical thermodynamics never really started. Indeed, matter in motion is inconceivable without the ideas of departure and arrival. The SLT is a law of divergence. It is like a travel schedule showing only departures. Its complement is a law of convergence. It is like a travel schedule showing only arrivals. Together, the SLT and its complement simply describe the motion of matter.

This modification of classical mechanics is consistent with the fundamentals of thermodynamics. For example, in the usual demonstration of the SLT (Fig. 6-1), chamber A is filled with gas and chamber B is essentially a vacuum. Opening the valve between the two (considered a "negligible" outside influence) allows gas from chamber A to enter chamber B spontaneously and irreversibly. This "spontaneity" is merely a reflection of the inertial motion of the gas molecules that, instead of colliding with the valve, now move through it. Entropy (or apparent disorder) increases as the molecules of the gas diverge from each other as they emerge from chamber A. The process is irreversible because all the gas molecules will not spontaneously return to chamber A. To produce a vacuum at chamber B and reestablish the previously "better-ordered" state, we would have to introduce some extremely significant outside influence clearly forbidden by the assumption that this is an isolated system. The strength of the classical view, not countermanded by Prigogine or anyone else, is its insistence that an ideally isolated system cannot, of itself, produce a net increase in order. **The source of the order producing mechanism must lie outside the system itself.**

---

[141] Nagel, *The Structure of Science*, p. 158.

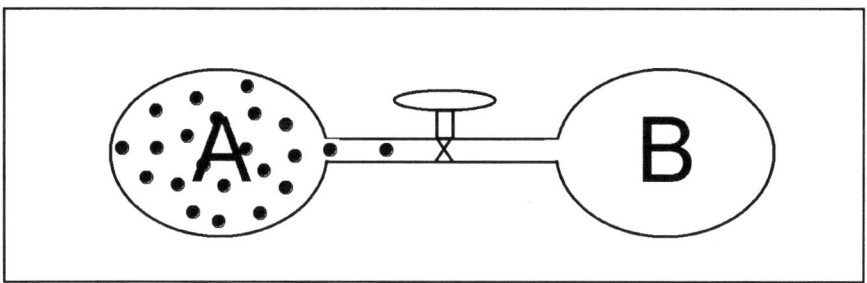

Fig. 6-1. The classical demonstration of entropy change described by the Second Law of Thermodynamics. An increase in entropy is produced when the gas in chamber A is allowed to pass through the valve into the vacuum of chamber B.

In this demonstration, the usual focus is on the divergence from chamber A, but if we view it from the perspective of chamber B, we see convergence instead. The gas molecules from chamber A rush in upon chamber B just as spontaneously and just as irreversibly as they left chamber A. If disorder has been produced in chamber A, order has been produced in chamber B. In an infinite universe, an increase in entropy in one place results in a simultaneous and equivalent decrease in entropy in another. The convergence of material entities results in an apparent increase in order or organization—the phenomenon that the SLT, by itself, cannot explain.

## SUBJECTIVITY OF ORDER-DISORDER

Until this point I have used "entropy" and "disorder" as fully interchangeable terms in line with the present convention. But entropy and disorder were not always considered interchangeable. When Rudolf Clausius introduced the term "entropy" more than a century ago, it was unclear what meaning, if any, could be assigned to it. At most, entropy simply meant "transformation."[142] Its association with disorder seems to have grown along with the acceptance of the Copenhagen interpretation of the Heisenberg Uncertainty Principle. Any objection to making entropy and disorder equivalent unavoidably must have a distinctive anti-Copenhagen flavor. For example, in information theory Claude Shannon drew his share of indeterministic criticism for considering entropy as the degree of ignorance about a system.[143] Entropy-ignorance-disorder: the implications of

---

[142] Tribus, Myron. "Entropy." In *The Encyclopedia of Physics*, edited by R.M. Besancon, 239-40. New York: Reinhold, 1966, p. 239.

[143] Shannon, Claude. "A Mathematical Theory of Communication." *Bell System Technical Journal* 27 (1948): 379-423, 623-56.

76 • The Ten Assumptions of Science

such a linkage are clear. The inclusion of an obviously subjective condition in the series, forces us into another look at the nature of the order-disorder concept.

Ordinarily there is little dispute that the classical demonstration of the SLT (Fig. 6-1) illustrates the production of disorder. But what can be said about changes in order when we modify the context of the demonstration as shown in Fig. 6-2? When the valve is opened, one can just as easily conclude that the system containing the 18 gas chambers becomes more orderly, not less so. Changes in entropy reflect objective changes in divergence or convergence, while order-disorder is purely subjective.

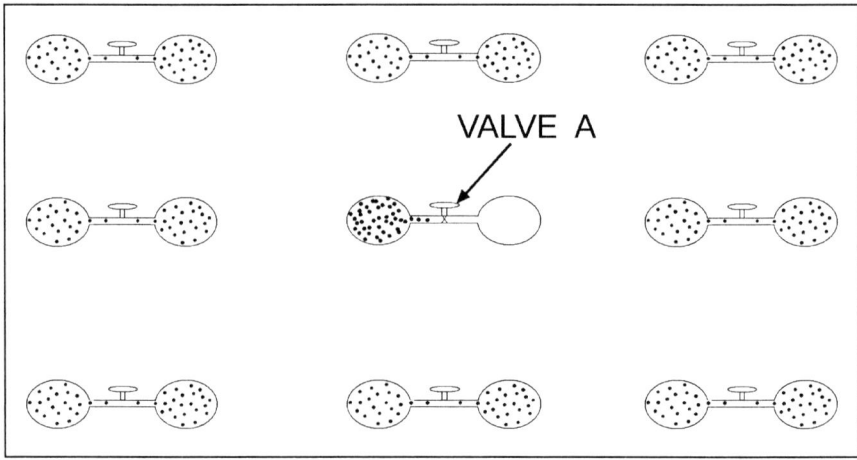

Fig. 6-2. Demonstration of the subjectivity of the order-disorder concept. Does turning valve A result in an increase or does it result in a decrease in order?

As much as this may upset our private feelings of order and disorder, it is nonetheless consistent with determinism, which considers the universe to be orderly, not disorderly; law like, not lawless. The upshot is that the SLT and its complement must describe something more than mere subjective changes in order.

## OBJECTIVITY OF DIVERGENCE-CONVERGENCE

As I pointed out before, the foremost assumption of mechanics is that the universe consists only of matter in motion. In mechanical terms, the SLT would be:

All bodies are subject to divergence from other bodies.

Its mechanical complement proposed here becomes:

All bodies are subject to convergence from other bodies.

The SLT-Order Paradox is resolved only by uniting thermodynamics with mechanics. In the process, the order-disorder concept necessarily loses objective meaning. Entropy becomes a statement about divergence, and its opposite, negentropy, becomes a statement about convergence. Subjectively, we can still view increasing disorder as things "fall apart" and increasing order as things "come together." Because the motions of matter are relative, the motion of a particular object may be a divergence for an observer at one point, while it may be a convergence for an observer at another point. Divergence and convergence are the essence of the motion of matter and must be considered objective and necessary features of the infinite universe.

The possibility of nearly ideal isolation derives from the possibility of divergence; the possibility of nearly ideal nonisolation derives from the possibility of convergence. In thermodynamic terms the complement to the SLT becomes: **the entropy or apparent disorder of an ideally nonisolated system can only decrease**. No object can be completely isolated, just as no object can be in an all-encompassing contact with its surroundings. Ideal isolation and ideal nonisolation are opposite ends of the continuum we use to describe the relationships between real objects and their surroundings.

With respect to each other, any two objects are semi-isolated to the degree of their separation and semi-nonisolated to the degree of their union. What we observe as increases in entropy for a particular system are results of the divergence of matter or the motion of matter from that system. What we observe as decreases in entropy for a particular system are results of the convergence of matter or the motion of matter upon that system. Isolation and nonisolation, therefore, are complementary aspects of the motion of matter.

# THE DIALECTICS OF MATTER IN MOTION

Because all matter in the universe is in constant motion, it is continually moving across or transmitting motion across system-environment boundaries. The entropy or state of divergence of a particular portion of the universe is always either increasing or decreasing. The SLT is a law of departure; its complement is a law of arrival. Ironically, the very ideal we required for formulating the SLT: perfect isolation, would prevent its operation. For the entropy of a system to increase, parts of that system must be able to interact with its environment.

To the degree that the system cannot transmit motion to the environment, it tends to expand, that is, it invades a portion of the universe formerly classified as "environment." Cosmogonists have applied this necessity for system expansion to the universe itself, but this is a *non sequitur*. The only requirement is for there to be an environment for the parts of a system to move into or to transfer motion to. An infinite universe in which matter and the motion of matter is not everywhere the same is sufficient.

The irreversibility to which the SLT and its complement speak is not a result of a grand, universal predominance of divergence over convergence, but simply a result of the motion of matter within an infinite universe. All systems, being in continual motion relative to each other, have a unique relation to all other systems in the universe at any moment. The motion of a system as a whole relates only to its surroundings. We must view the apparent production, maintenance, and destruction of order, not as a property of the system, but as a relationship between system and environment.

The question arises as to the experimental relevance of this mechanical complement to the SLT. We will continue to study the interactions of subsystems in which entropy (or disorder, from the subjective point of view) is produced and destroyed as subsystems diverge and converge. Nevertheless, because subsystems are always parts of larger systems and these are parts of still larger systems, we must expect eventual convergence from systems unfamiliar to us. The complement to the SLT, convergence, ultimately must be a law of the unknown—a law that predicts that no matter how much we widen the boundaries of a system, there will always be matter in motion outside that system.

The philosophical shift from the system-oriented approach to the system-environment approach resolves the SLT-Order Paradox. The acceptance of COMPLEMENTARITY for the Second Law of Thermodynamics requires an acceptance of the other assumptions of science. **Noncomplementarity**, the indeterministic alternative, can exist only in a finite universe in which the system is considered more important than its environment. The rejection of this "system myopia" will be the culmination of the great work that Copernicus began.[144]

---

[144] For those who might enjoy it, I include the mathematical formulation of COMPLEMENTARITY:

In mathematical terms the SLT is:

$$dS > dQ/T \qquad (6\text{-}1)$$

Where:
$dS$ = change in entropy of the system
$dQ$ = change in motion within the system
$T$ = temperature, degrees Kelvin

By symmetry, the complement to the SLT is:

$$dZ > dO/T \qquad (6\text{-}2)$$

Where:
$dZ$ = change in entropy of the environment
$dO$ = change in motion within the environment

CHAPTER 7

# THE SEVENTH ASSUMPTION OF SCIENCE:
# IRREVERSIBILITY

*All processes are irreversible.*

We are continually reminded—often sadly—of significant features of our lives that are unmistakably irreversible: we lose teeth, hair, and other body parts. Friends grow old and die. Instinctively we know that there is no traveling back in time although we may dream of it and try to recreate the conditions of a former happy period.

We feel the passage of time described by the Seventh Assumption of Science, IRREVERSIBILITY. And what **is** time? Does time occur independently of matter? Is it a property of matter? Is it another dimension of matter? Is it a concept? Is it a measurement? Or is it as Santayana said, just "another name for the native instability of matter"?[145]

Time is a perennial subject of philosophical contention. In this debate it is the special mission of indeterminists to portray time as an unfathomable mystery, while it is the business of determinists to portray time as inseparable from matter.[146] As we were encouraged to do under INSEPARABILITY, the first step in the analysis of any phenomenon is to determine whether that phenomenon is matter or whether it is the motion of matter. Even to the naive observer, time is clearly not matter. It certainly does not exist in the way that matter exists. Although we may speak of finding a "chunk of time" to carry out one of our special projects, only a fool would search for it in a literal sense. Time does not "exist," it "occurs."

---

[145] Santayana, George. *The Realm of Matter*. London: Constable, 1930, p. 83.
[146] Coe, Lee. "The Nature of Time." *American Journal of Physics* 37 (1969): 810-15.

Next we must distinguish between time and the concept of time. The problem is similar to that discussed under MATERIALISM. Matter, we assumed, exists external to us. Similarly, time occurs external to us. The concept of matter was an abstraction we used when referring to "all things." We assumed that matter *per se* does not exist, that only particular examples of matter exist. Likewise, the concept of time is an abstraction we use when referring to "all events." Strictly speaking, time *per se* does not occur, only particular events occur. As with the existence of matter, time occurs independently of us regardless of what we are able to say about it.

In practice, the concept of time encourages the attempt to relate the motions of one thing to the motions of other things. Time is thus an echo of CAUSALITY. The concept of time teaches us to relate the specific to the general, to view a system in its relation to the rest of the universe. The belief in IRREVERSIBILITY requires an acknowledgment of the importance of the environment, while its indeterministic alternative, **reversibility**, habitually denies it.[147]

In asserting that some processes are reversible, indeterminists reflect not only the narrow perspective of systems philosophy, but also the dreams of political reactionaries who seek to "turn back the hands of time." Some form of the belief in **reversibility** seems likely to remain with us indefinitely even though it becomes more and more untenable each day. Planck was right in pointing out that the antithesis between IRREVERSIBILITY and **reversibility** that he thought irresolvable would play a leading role in the development of the scientific worldview.[148]

## HISTORY OF IRREVERSIBILITY

A feeling for IRREVERSIBILITY develops in anyone who has ever watched a birth, a death, a beginning, or an ending. It is a basic theme of poetry, art, and philosophy. In science though, it did not receive its greatest impetus until the antiquity of the universe first became apparent.[149] According to Stephen Toulmin: "This 'discovery of time' has taken place almost entirely since A.D. 1800, and the midnineteenth century debates about evolution were only one small but particularly noisy episode in a much larger intellectual revolution."[150]

---

[147] Conger, G.P. *Synoptic Naturalism*. Minneapolis: University of Minnesota Library, 1960, p. 197.

[148] Cassirer, *Determinism and Indeterminism*, p. 80.

[149] Toulmin, Stephen. "The Discovery of Time." *Manchester Literary and Philosophical Society* 105 (1962): 100-12; Toulmin, Stephen, and June Goodfield. *The Discovery of Time*. New York: Harper, 1965.

[150] Toulmin, "*The Discovery of Time*," p. 100.

Laplace is typical of the many scientists who "discovered time" in the course of their work. He had spent the first quarter of the century trying to prove that the solar system is eternally stable, that it has always existed. But then, almost as an afterthought, he included a footnote in which he demolished his main point by proposing an origin for the solar system. From thenceforth the development of the solar system was to be seen as irreversible.

In geology too, the same conservative instincts were at work when Hutton popularized the uniformitarian principle.[151]. Uniformitarianism declared that the same motions were repeated over and over again. The past was the perfect key to the future. Taken literally and absolutely, uniformitarianism, like Laplacian determinism, was just another version of finite universal causality. There could be no progression. Of course, the fossil evidence noted by Cuvier in 1810, belied this. Used at first in support of uniformitarianism (defined at that time as slow natural change) in the battle against catastrophism (defined at that time as rapid supernatural change), fossil progression implied that the past was a less than perfect key to the future. The unprecedented elements in the historical record signaled the beginning of the end for finite universal causality and **reversibility**.

In biology[152] and sociology[153] IRREVERSIBILITY also made huge inroads. But it was chemistry that came closest to establishing it as a fundamental law of the universe. As we have already seen, the rejection of the caloric theory in the 1840's forced the concept of motion to be included along with the concept of matter in the assumption of CONSERVATION (First Law of Thermodynamics). Further development in this new subdiscipline of chemistry could no longer avoid IRREVERSIBILITY. The Second Law of Thermodynamics (SLT), became, in effect, the discovery of chemical evolution.[154] And as Lewis and Randall pointed out, the SLT encountered considerably more philosophical resistance than the First Law of Thermodynamics.[155] As long as the First Law

---

[151] Hutton, James. "The System of the Earth, Its Duration, and Stability." In *Philosophy of Geohistory: 1785-1970*, edited by Jr. Albritton, C.C., 24-52. Stroudsburg, PA: Dowden Hutchinson & Ross, 1785 [1975].

[152] Darwin, Charles. *The Origin of Species by Means of Natural Selection or the Preservation of Favoured Races in the Struggle for Life.* 6 ed. New York: New American Library, 1859.

[153] Marx, Karl. *Capital: A Critique of Political Economy.* Vol. 1. New York: Vintage, 1867.

[154] Clausius, Rudolf. "English Translation of Pogg. Ann. 93:481-506." In *Reflections on the Motive Power of Fire*, by Sadi Carnot, and Other Papers on the Second Law of Thermodynamics by E. Clapeyron and R. Clausius, edited by E. Mendoza, 152. New York: Dover, 1854.

[155] Lewis and Randall, *Thermodynamics.* 1923, p. 110.

was interpreted only as a statement about the conservation of matter it could be construed as being compatible with **creation**. For many scientists, the idea of supernaturally created matter accorded with the idea of naturally unchanging matter. Religious-atomistic views were threatened as soon as matter was discovered to be naturally changing and without the permanence assumed by **certainty** and **separability**.

As the static view gradually gave way to the dynamic, believers in **reversibility** developed a compromise that acknowledged the occurrence of motion, but only in an offhand way. Grudgingly, they admitted that the motions of things within isolated systems went forward, but were quick to point out that they went "backward" too. For idealists, the perfect repetition of the internal motions of the isolated system could produce perfect equilibrium—a new kind of perfect rest. Such a system, displaying no real change over time, was, like the perfectly solid matter of the atomist, a bulwark against the idea of evolution.

It is ironic that the fundamentally conservative idea of isolation conspired against its masters in the development of the SLT. At first, systems supposed to be free from outside interference were viewed as possessing cyclic inner motions capable of continuing indefinitely. Such systems would be "perpetual motion machines," and since they never exchanged matter or motion with their surroundings, they would exist forever in their originally "created" forms. This vision was soon found to be hopelessly unrealistic. Even those systems coming closest to being perfectly isolated were only approximately so. Every system-environment interface anyone could design was "leaky." Matter invariably got through the "holes" in the containers or transmitted some of its motion through the walls. Worst of all, without continuous contact with an environment identical to the one that produced it, every system underwent an irreversible dissipation.

As already mentioned under COMPLEMENTARITY, the systems interpretation of the SLT foretold a strange kind of "progress." It was one in which things tended to fall apart rather than come together. It spoke of increasing destruction and disorder rather than increasing construction and order. Without its complement, interpretations of the SLT were invariably pessimistic and regressive. Such views were very much a reaction to the trials and tribulations of the Industrial Revolution.[156] Even so, with the SLT, systems philosophy had destroyed forever the myths of permanency and **reversibility**. It did not require an acceptance of progress to do it.

---

[156] Rifkin, *Entropy*.

# HOW COMPLEMENTARITY IMPLIES IRREVERSIBILITY

Even without its complement, the SLT gave modest support to IRREVERSIBILITY. Certain leaders in classical thermodynamics, realizing that perfect isolation was only a fiction, were aptly dogmatic: "any actual process is said to be **irreversible**."[157] No matter what the system, the SLT correctly predicted that eventually it would lose matter and motion and that it would not, by itself, be able to restore that matter and motion. There was no doubt that without **reversibility** the universe could not be both eternal **and** finite.[158]

## The Necessity for an Infinite Universe

As explained under COMPLEMENTARITY, each portion of the infinite universe is in motion with respect to all the other portions. The departure of matter and motion from one portion of the universe always implies an arrival of that matter and motion at another portion of the universe. In such a universe, divergence is equivalent to convergence. The SLT becomes a law of divergence and its complement becomes a law of convergence.

That would not be true for a finite universe. In a finite system, say three items confined to a table top, one can consider the equivalence of divergence and convergence instead as a support for **reversibility**. Any of the items can be moved around the surface of the table and then returned to its previous position. By sticking closely to systems philosophy, even the convergence produced by the hand that moves the item could be viewed either as insignificant or as perpetually repeatable. Any finite number of items and any finite portion of the universe would produce the same result: a finite number of combinations. Such a finite system does not evolve, because nothing new is allowed to enter it. It can only repeat itself, endlessly producing and destroying the same limited set of forms.

So IRREVERSIBILITY cannot be justified from a strictly system-oriented point of view. The SLT was successful only because, ever so slightly, it was forced to admit the existence of the environment. Without this "dumping ground" for the matter and motion that left it, the system would have been everlasting and the phenomena described by the SLT could not occur. In an infinite universe of objects in continuous motion, the system-environment relationship of each object is continually changing. The parts of a system, such as the table top with the three items on it, considered in isolation from the rest of the universe, may be

---

[157] Lewis and Randall, *Thermodynamics*. 1923, p. 112.
[158] Elliot, Hugh. *Modern Science and Materialism*. London: Longmans Green, 1919, p. 61.

placed in successive, identical relationships with each other. But as explained under CAUSALITY, the relationship between any two objects is never independent of relations with still other objects, be they within the system or without. At the same moment that any two objects seem to be approaching a former relationship, other objects in the infinite universe are converging on them and diverging from them, ensuring that the relationships between the two objects and others outside the system are never identical at subsequent moments. Contrary to the indeterminist,[159] this does not negate causality—only the finite form of it. As Santayana so wisely put it: "All movements of matter are…responsive afresh to a total environment never exactly repeated, so that no single law would perfectly define all consecutive changes, …every response would be that of a newborn organism to an unprecedented world."[160]

CAUSALITY, UNCERTAINTY, and IRREVERSIBILITY thus are consuponible. In other words, if one assumes that all effects have an infinite number of causes, then it is also necessary to assume that an effect will never occur in **exactly** the same way twice. Not only are any causal laws we can devise finite and therefore incomplete, but they also are derived from previously occurring causes. They cannot have all of the new elements that will contribute to a similar effect in the future.

# THE MYTH OF REVERSIBILITY

**Reversibility** remains a viable indeterministic alternative to IRREVERSIBILITY. But, just as few people admit to an outright belief in **acausality**, few admit to an outright belief in **reversibility**. Nonetheless, **reversibility** often receives credence in specific instances. For example, it is not uncommon to find scientists who consider movements in opposite directions at equilibrium to be indications of **reversibility**. Even Max Planck believed that "the concept of entropy had a physical significance only where there could be a reversible process."[161] Viewed in a limited way, that is, from the point of view of systems philosophy, phenomena such as gravitation, mechanical and electrical oscillations, sound waves, and electromagnetic waves are commonly considered reversible in an absolute sense.[162] The correct interpretation though, is that it is IRREVERSIBILITY that is absolute and **reversibility** that is not. The arrow of time points in only one direction.

---

[159] Cassirer, *Determinism and Indeterminism*, p. 58.
[160] Santayana, *The Realm of Matter*, p. 110.
[161] Planck, *Where Is Science Going?* p. 185.
[162] Ibid., p. 189.

## Microscopic "Reversibility"

The challenge to IRREVERSIBILITY is, like **acausality**, inserted into scientific discourse in subtle ways, if not inadvertently. For example, in *Geomorphology and Time,* Thornes and Brunsden were moved to write: "time is distinguished by possessing the property of intrinsic direction and in the **macroscopic** sense being irreversible [emphasis mine]."[163]

The implication, of course, is that **reversibility**, like **acausality**, is probably rare, but might be possible in the microscopic realm. The argument for **reversibility** must be squeezed into the same small philosophical space occupied by the assumptions of **certainty**, **separability**, and **acausality**. Primarily, it requires the estrangement of the concept of motion (i.e., time) from CAUSALITY. It depends on the limitations of direct observation and on the necessity for probabilistic laws as a sort of grand *argumentum ad ignorantiam*.

As long as this kin of the Copenhagen viewpoint remains dominant, microscopic **reversibility** will have its overt defenders[164] and time cannot be viewed properly—as motion. Typically the confusion[165] is manifest at the point where systems philosophy breaks down: where the claims for ideal isolation obviously no longer fit the reality.[166] As subsystems approach equilibrium with each other, the relationship between the system and its environment becomes increasingly important. By continuing to ignore the environment as equilibrium develops, one must invent evermore overtly indeterministic interpretations of the phenomena under investigation.[167] At the extreme, the so-called reversible reactions near equilibrium have been thought to engender "cyclic time" which supposedly involves a sort of "time-canceling effect."[168]

Finally, in the words of our 20th century indeterminists, "the situation in steady conditions is…time independent"[169] and "the whole point of the analysis of measurement on the microscopic level (is) that there is nothing to abstract

---

[163] Thornes, J.B., and D. Brunsden. *Geomorphology and Time.* New York: Halsted Press, 1977, p. 2.

[164] Rosnay, *The Macroscope*, p. 162.

[165] Prigogine, "Time, Structure, and Fluctuations," p. 785.

[166] Blum, H.F. *Time's Arrow and Evolution.* Princeton, NJ: Princeton University Press, 1968, p. 109.

[167] Gardner, Martin. "Can Time Go Backward?" *Scientific American* 216, no. 1 (1967): 98-108; Rosnay, *The Macroscope*, p. 168; Laszlo, Ervin. *The Systems View of the World: The Natural Philosophy of the New Developments in the Sciences.* New York: Braziller, 1972, p. 52.

[168] Thornes and Brunsden. *Geomorphology and Time,* p. 2.

[169] Ibid., p. 140.

from."[170] This application of **reversibility** goes a long way in fulfilling the ancient desire of the indeterminist. To be independent of time would be for matter to be without motion, at absolute rest, isolated from the rest of the universe, independent from other matter in motion. The idea of time independence ultimately promises life eternal since it calls for matter to sit still and ultimately to disappear—the solipsist's dream.

## DOES CAUSALITY REQUIRE REVERSIBILITY?

To some, it may appear that without **reversibility**, perhaps even on a macroscopic scale, causality would be impossible. For successful prediction, wouldn't it seem that events must have some chance of recurring? In a previous passage Santayana eloquently expressed the uniqueness of every event, but here he takes an apparently opposed view: "Whatsoever spontaneously happens once will have spontaneously happened before and will spontaneously happen again, wherever similar events are in the same relation."[171]

The answer to this apparent contradiction appears in a single word: "similar." All events are "similar" and not "identical" as would be required in an absolute conception of reversibility and the finite conception of causality against which I have argued. Only a conception of CAUSALITY as infinite is in accord with IRREVERSIBILITY and the notion of "universal time," the feeling that every motion of every part of the universe is unique and never will be repeated.

### Time Independence?

If time is motion, as I have assumed, then according to INSEPARABILITY, it cannot exist apart from matter. As we have seen, modern indeterminists, of course, would not be up to date if they did not in some way hypothesize the **separability** of matter and motion. As explained under CONSERVATION, it is a special preoccupation of certain indeterminists to try to imagine a "time" that "existed" "prior" to the existence of the universe. Similarly, the hypothesis of microscopic reversibility rests on the **separability** of matter and motion—taken here as time independence. Although even modern physics teaches time dependency, this is in daily confrontation with indeterministic views advocating time independence. The clash represented by these opposed positions still remains too confusing and unnerving for those who are not quite sure and are all too quick to

---

[170] Hawkins, David. "A Causal Interpretation of Probability." Ph.D. thesis, University of California, 1940, p. 174.

[171] Santayana, *The Realm of Matter*, p. 109.

make the compromise between science and nonscience: "It is not the purpose of this book to perpetuate the rather fruitless polarization of the subject between...advocates of time dependency and time independency."[172]

Of course, the polarization is a reflection of the perpetual, progressive conflict between determinism and indeterminism. Only an indeterminist could regard the debate as fruitless. This typical, timorous approach, characteristic of Western attempts at scientific education hardly can be conducive to efficient scientific development.

There is no time independence. This is first grade stuff and there is no point in being indecisive about it. Ever since Einstein's Special Theory of Relativity, the idea of time independence has been defunct. Remember, under INSEPARABILITY we considered space as matter and time as motion. Even Minkowski hinted at this in his famous statement of 1908: "Henceforth space by itself, and time by itself, are doomed to fade away into mere shadows, and only a kind of union of the two will preserve an independent reality."[173]

# IRREVERSIBILITY AND THE ENVIRONMENT OF THE SYSTEM

In the deterministic view, absolutely no reaction or event "occurs" independent of time because time is motion; the terms "reaction" and "event" are terms describing motion. No reaction or event is reversible in the absolute sense because the material objects in the environment in which it takes place are in continuous motion. Each star, each galaxy moves relative to all the others. The night sky is unique each time we view it. Reversibility could only occur in systems that are completely isolated from the rest of the universe. Such isolation is impossible and therefore **reversibility** is impossible. The Seventh Assumption of Science, IRREVERSIBILITY, asserts that **all** processes are irreversible. We need not delude ourselves into thinking that there are some exceptions involving ideal equilibrium or relative size.

---

[172] Thornes and Brunsden. *Geomorphology and Time*, p. 2.

[173] Minkowski, H. "Space and Time." In *The Principle of Relativity*, edited by A. Einstein, H.A. Lorentz, H. Weyl and H. Minkowski, 75-91. New York: Dover, 1908 [1923].

# CHAPTER 8

# THE EIGHTH ASSUMPTION OF SCIENCE: INFINITY

*The universe is infinite, both in the microscopic and the macroscopic directions.*

The universe is either infinite or finite. It is, of course, impossible to **know** for sure which of these possibilities really exists; we can only **assume** one or the other. Whenever someone claims to have found the edge of the universe, someone else is free to hypothesize objects beyond. Experience in particle physics and astronomy teaches us the utility of the Eighth Assumption of Science. And yet, each time a supposed limit is reached, the absence of data allows a choice between INFINITY and **finity**, its opposite.

As I have mentioned, two very different scientific approaches have vied for the title of the "scientific worldview": classical mechanism and systems philosophy. Classical mechanism emphasized the external interactions of its model; systems philosophy emphasized the internal interactions of its model. In the evolution of the two points of view, the invention of the glass lens was both symbolic as well as instrumental. If the telescope was the tool of classical mechanism, the microscope was the tool of systems philosophy. The more one saw of the macroscopic world, the more one was impressed by its immensity; the more one saw of the microscopic world, the more one was impressed by its inexhaustibility. Unfortunately, the more one looks in one direction, the more one learns about that direction—and forgets about the other. Classical mechanism and systems philosophy developed two different views of infinity: macroscopic and microscopic.

# MACROSCOPIC INFINITY AND CLASSICAL MECHANISM

Although philosophy contains numerous allusions to the possibility that the universe is infinite in both the macroscopic and microscopic directions, this always has been a heretical view. The macroscopic infinity proposed by Democritus and other atomists gave way during the Middle Ages to the practical demands of astronomy and religion. In tune with the widespread techniques developed in oceanic navigation, Ptolemaic cosmology invented the theory of the two-sphere universe in which the fixed sphere of the Earth was surrounded by a moving celestial sphere. This explained the obvious rotation of the night sky about the North Star and allowed for the fantastic heaven beyond. Thus when Giordano Bruno resurrected the theory of an infinite universe, he was, in effect, challenging the notion of an actual physical existence for paradise. This would not do. The reaction by the church was swift and terrible. Eschewing a Galilean-type recantation, Bruno was burned at the stake in 1600.

Nine years later, Galileo's telescopic observations accelerated an enumeration remarkable for its unceasing ability to furnish evidence for Bruno's speculation. With the telescope one could study the interactions of astronomical bodies. One need not be bothered particularly by questions concerning the constituents of those bodies. Thus, it was logical that Newton's later development of mechanics would pursue this obsession with the external as opposed to the internal. Like Bruno's less pious version of infinity, Newton's was macroscopic, not microscopic. Theirs was a vision of an infinite number of solid bodies in an infinite volume. Both retained the legacy of atomism, which, without a doubt, presumed microscopic finity.

Classical mechanism made its compromise with finity in the traditional way: by neglecting, and thereby finalizing and finitizing the insides of its model. Thus when Newtonian bodies collided, the irreversible effects on the internal motions of those bodies were not understood and had to be ignored. Newtonian bodies had permanence consistent with conservative, nonevolutionary views of society. Classical mechanism was a bitter pill, but it could be swallowed.

# MICROSCOPIC INFINITY AND SYSTEMS PHILOSOPHY

For more than half a century, almost no one thought to turn Galileo's telescope around and look inward instead of outward. Then, in 1665, Robert

Hooke's observations of the cellular structure of cork initiated a search, which, like Galileo's, has never failed to provide evidence for microscopic infinity. The pace of discovery in this direction, however, continued to lag behind that of the macroscopic. For instance, as early as 1683, Anton van Leeuwenhoek reported some casual observations of what only could have been bacteria, but these were not to be observed again for at least another century. A worldview that included microscopic infinity among its basic assumptions was to develop alongside classical mechanism, but it was to remain in the background until it could manufacture its own special compromise with **finity**.

Indeterminism showed its great strength by disallowing the straightforward conclusion that the data from the telescope and the microscope warranted an assumption encompassing infinity in both directions. It was left to systems philosophy, present in one form or another ever since the evolution of the first ego, to develop the only choice possible under such circumstances. Ever adept at overemphasizing the internal as opposed to the external, systems philosophers gradually overcame the tremendous lead of the classical mechanists, who overemphasized the external as opposed to the internal. The prospects for microscopic infinity grew while those for macroscopic infinity withered.

Inevitably, the universe, too, would again be viewed from a system-oriented perspective. The whole universe, even though known to be billions of light years in extent, was to be viewed in isolation: a system without an environment. Twentieth century science admitted that systems evolved, but as I have explained under COMPLEMENTARITY, it was reluctant to acknowledge the full impact of the environment. From the system-oriented point of view, the universe **had** to be finite. At the same time that science was leaning toward microscopic infinity, indeterminism was pressing for a return to macroscopic finity. Systems philosophy found the way.

To understand why systems philosophy was so successful in preparing the accommodation, we must recall the great theoretical revolution initiated by Heisenberg. From Democritus to Laplace it was possible to advance the cause of determinism by assuming microscopic finity along with macroscopic infinity. But as we have seen, the Laplacian form of determinism is clearly unworkable. Its underlying assumption of **finity** is no longer compatible with the other Assumptions of Science. No longer could classical mechanism, based on this form of determinism, seriously propose that the ultimate constituents of matter eventually could be described in full. No longer was there any hope of realizing Einstein's dream of discovering the elementary laws by which the universe could be built up by simple deduction. According to CAUSALITY and UNCERTAINTY, partial descriptions and fallible predictions are the only attainable reality.

As mentioned, through the Copenhagen interpretation, systems philosophy was able to distort the true meaning of the overthrow of Laplacian determinism. It did this by borrowing from classical mechanism the notion that a complete description was possible: one only had to consider chance as a singular cause. Systems philosophy was thus ambivalent even toward microscopic infinity. It could refuse to take Bohm's "subquantic states" seriously, even as it led the search for smaller and smaller particles. Systems philosophy leaned toward microscopic infinity just enough to switch theoretical attention from the external to the internal and just long enough to overthrow macroscopic infinity. The new focus was quite sufficient for the crowning achievement of systems philosophy: the Big Bang Theory of the origin of the universe.

The ambiguities introduced by the modern attempt to make microscopic infinity compatible with macroscopic finity are no easier to live with than the ones devised by classical mechanism to do the reverse. Both approaches ultimately ended up supporting a grand **finity** even in the areas in which they chose to specialize. In actuality, of course, microscopic infinity logically implies macroscopic infinity and vice versa. This follows from many of the previous discussions, particularly the one on "spacetime" involving the opposed concepts of ideal "solid matter" and ideal "empty space." Again, if we grant these to be merely idealizations, then we are assuming that they have no actual existence. Neither indivisible matter nor immaterial void are possible, although each portion of the universe reflects both to varying degrees. Without actually existing indivisible matter there can be no microscopic finity just as there can be no macroscopic finity without an actually "existing" immaterial void outside the universe.

We can no longer merely entertain either microscopic or macroscopic infinity; we must steadfastly assume both. The resulting assumption of INFINITY, the proposition that **the universe is infinite both in the microscopic and the macroscopic directions**, is the only form compatible with CAUSALITY and UNCERTAINTY. Such an assumption is not to be taken lightly, as the struggle over CAUSALITY and UNCERTAINTY indicates.

There is, of course, an unavoidable, necessary circularity to the argument for INFINITY. This is a characteristic to be expected if the universe was truly infinite and assumptions rather than absolute presuppositions or *a priori* were necessary for its study. Because it includes the macroscopic as well as the microscopic, INFINITY stands out so strikingly different from the currently accepted view that it leads, as we have seen, to a major reinterpretation of all our scientific assumptions. Now let us examine some of the factual evidence upon which the belief in INFINITY rests.

# QUEST FOR THE ULTIMATE PARTICLE

The ultimate particle has been the object of a continuous search ever since Democritus proposed it 2500 years ago. In his view, the ultimate constituents of matter were atoms, hard little balls filled with an inert substance called "matter." These atoms were unprecedented in that, unlike other things, they could not be further subdivided. Also unlike other things, all atoms had identical properties.

The atomic concept, of course, was useful; matter **does** appear to consist of discrete particles. The atom was considered the basic building block of matter well into the 19th century before its supposed unchanging nature came under question. As more and more information on atoms was accumulated it became obvious that all atoms could not be considered identical as the atomists' assumption of microscopic finity demanded. True, the basic properties of the constituent atoms of a metal such as gold appeared identical, but the atoms of gold and the atoms of other elements differed in many ways. For example, a million atoms of gold did not weigh the same as a million atoms of silver. How could the atom be the ultimate particle if different atoms had different masses? There must be something inside these atoms—perhaps this something was the ultimate particle instead.

Sure enough, by 1897 J. J. Thomson had discovered the electron, proving beyond a doubt that the atom was divisible. This began a long series of discoveries that alternately raised the possibility of microscopic infinity whenever a subatomic particle was split, and lowered it whenever there was a failure to split. To this day, each subatomic particle, when explored with some newly invented technique, eventually yields still another, even smaller particle.

The electron has been succeeded by the quark as the smallest subatomic particle.[174] Will there be an end to this succession? Will an ultimate particle be found? One scientist felt sure that "matter is not infinitely divisible,"[175] while another reiterated that "no 'ultimate' individual or partless particle is known to science."[176] Indeterminists once asserted that "the electron does not have other particles inside it,"[177] while determinists just as dogmatically asserted that the electron was just as inexhaustible as the atom.

---

[174] Barnett, R. Michael, Henry Muehry, and Helen R. Quinn. *The Charm of Strange Quarks: Mysteries and Revolutions of Particle Physics*. New York, NY: American Institute of Physics and Springer-Verlag, 2000.

[175] Dobzhansky, Theodosius. *Mankind Evolving: The Evolution of the Human Species*. New Haven, CT: Yale University Press, 1962, p. 25.

[176] Parsons, H.L., ed. *Marx and Engels on Ecology*. Westport, CT: Greenwood Press, 1977, p. 5.

[177] Perl, Martin. "Leptons—What Are They?" *New Scientist* 81 (1979): 564-66, p. 564.

In the specific, determinists won out: there are now at least three different kinds of electrons, some of which emit neutrinos.[178] But who would be so foolish as to declare that this will always be the case? There can be no end to the debate between those who believe in microscopic infinity and those who do not. There is no experiment that could settle the question once and for all.

## LOOKING FOR THE EDGE OF THE UNIVERSE

As we grow up, we find that there are no limits to our environment. Every door opens on another, and another, and another. The intellectual growth of humanity has followed a similar pattern. Our acknowledged environment has expanded from the village, to the continent, to the Earth, to the galaxy and beyond. Before the 1920's we thought we lived in an island universe situated within an infinite volume completely void of matter. But then, the discovery of other galaxies—over 100 billion at last guess—once again dashed indeterministic hopes that the environment was finite. If there was literally something "beyond physics," it had to be at least $10^{23}$ kilometers away and impossible to reach in less than 13.7 billion light years. Each time that its vision dimmed, humanity faced the ancient metaphysical choice: either what is beyond physics is physical or it is not.

Today that choice requires a leap of faith just as much as it did in the past. In some ways our current ignorance of what lies beyond the reach of the largest telescopes is fundamentally no different from that of the pre-Columbian Europeans who were unaware of the New World. From hindsight, of course, it is easy to point out that those who assumed that the environment was infinite in the macroscopic direction were scientifically correct. But is this likely to be true in the future? We can never know for sure; the widespread prevalence of the opposing assumption is proof of that.

To avoid macroscopic infinity, indeterminists today follow one of two paths. The first visualizes the known universe situated within, and perhaps expanding into, an infinite immaterial void. The second seeks refuge in the mathematics of Einstein's 4-dimensional, finite universe. The simple deterministic alternative to such whimsy requires neither an immaterial void nor an unprecedented geometry.

## INFINITY: MICROSCOPIC PLUS MACROSCOPIC

As mentioned, classical mechanism emphasized macroscopic infinity and systems philosophy emphasizes microscopic infinity. Regardless of the special pleas

---

[178] Weisskopf, V.F. "Contemporary Frontiers in Physics." *Science* 203 (1979): 240-44, p. 243.

one can make for one exclusive of the other, the two ideas logically imply one another. The indeterministic notion of the ultimate particle with no thing (e.g. nothing) inside it is of a sort with the indeterministic notion of a universe with no thing (e.g. nothing) outside it. The philosophical purpose of **finity**, whether it be construed as microscopic, macroscopic, or both, is at some point to call a halt to MATERIALISM. On the other hand, the philosophical purpose of INFINITY is to proclaim the universality of MATERIALISM.

## Describing INFINITY

Those who support the concept of INFINITY are often asked to describe or even to define it. But of course a complete description of INFINITY would be a contradiction in terms: "As the universe is infinitely varied, it is very likely that only statements of infinite length can be true...the ontological structure of the universe is such that all universal statements of finite length **are** false."[179]

Nevertheless, most any child can begin a model of infinity by stacking blocks in all directions. Any attempt to enumerate the members of a class or to repeat a process endlessly amounts to a primitive model of infinity. The method of simple enumeration, for example, was used by Gamow[180] in retelling the story about the hotel with an infinite number of rooms. Whenever a new guest arrives, the previous arrival makes space available by moving into his predecessor's room, who, in turn, moves into his predecessor's room. Each guest advances one room each time a new guest arrives. Thus, the infinite hotel with the infinite number of rooms can always accommodate an infinite number of guests.

Planck described infinite time by comparing it to the cooling of a hot iron in water: "The smaller the difference of temperature between the hot iron and the water the slower is the transmission of heat from the one to the other, and calculation shows that an infinitely long time passes before an equal temperature is reached. This means that there is always some difference of temperature no matter how much time is allowed to elapse."[181]

As models of infinity, these are typical of the system-oriented viewpoint. They are inadequate because, even though they attempt to model processes, the processes invariably occur in isolation and thereby develop a static nature. The same event is repeated endlessly. These descriptions are dependent on the assumption of **reversibility**. Planck's pseudoscientific illustration is not much better than

---

[179] Lakatos, Imre, John Worrall, and Gregory Currie. *Mathematics, Science and Epistemology*. Vol. 2. New York: Cambridge University Press, 1978, p. 123.

[180] Gamow, George. *The Creation of the Universe*. New York: Viking, 1961.

[181] Planck, *Where Is Science Going?* p. 189.

Gamow's quixotic one. Both amount to a negation of INFINITY because, for the demonstration to proceed, it must occur in isolation.

The essence of INFINITY, on the other hand, lies in the possibility of relative nonisolation: convergence from the environment. In Planck's example neither the hot iron nor the water used to cool it could escape the impingement or the diminished impingement from the infinite number of things that lie outside of it. Planck first imagines and then calculates that the water will always be cooler than the iron, but the actual cooling of an iron must occur in a real environment that has a real temperature. As the temperature of the iron approaches that of the water, the impact of even one atom or photon from this environment could increase the vibrational motion of one of the water molecules enough to make the temperature of the water higher than that of the iron. The result: end of experiment; end of within-class infinity.

The mathematics that Planck alluded to was no proof of INFINITY, just as Plato's ideal geometric forms were no proof of their real existence. The success of mathematics as well as of science depends on its willingness to assume, in other words, to make a beginning. Only by divorcing itself temporarily from the infinite external world can mathematics or science reach a conclusion. Just as the real world intruded upon Plato's idealism, so does it intrude upon the necessarily finite mathematical physics of today. But try as we may, INFINITY won't go away: "The infinities that occur in QED (quantum electrodynamics) are clearly symptomatic of some profound shortcomings in our understanding of physics."[182] You bet.

These shortcomings, of course, are none other than those of systems philosophy itself. Our infinite universe refuses to be jammed into the finite, isolated container we have contrived for it. As explained under UNCERTAINTY, no theory can be complete because any particular class of things is susceptible to interference from other classes of things. Whether we recognize our ignorance with formal mathematical symbols or in some other way, we still must recognize it. We invariably give up contact with the real world whenever we use mathematical axioms that "somehow avoid the concept of infinity."[183]

When it deals in specifics and ignores the rest of the universe, mathematics forces thought into the finite mode. When it deals in generalities and slips from its disciplined course, mathematics regularly runs into the notion of INFINITY. Well-trained minds of the 21st century persist in the grooves provided by systems philosophy. Subconsciously they know that if the universe is truly finite, as the cosmogonists say it is, then the mere thought of INFINITY also has no place in mathematics.

---

[182] Davies, Paul. "Infinite Problems of the Very Small." *New Scientist* 83 (1979): 284-86, p. 286.

[183] Edmonds, "Can a World without Infinity Be Compatible with the Real Numbers?" p. 83.

Mathematicians nevertheless have persisted with such thinking despite cosmogonical fad. The development of the "nonstandard model" of infinity by Abraham Robinson is a case in point.[184] Previous models were based on a one-dimensional view of the "standard infinite line": the line one can imagine extending forward and backward for an infinite distance and consisting of an infinite number of segments. When the third dimension is considered, the picture changes drastically. One then imagines an infinite number of lines radiating from any point or segment of the standard infinite line. Each of these lines is, in turn, an exact copy of the original line, having infinite length and an infinite number of segments. The result is a "complex structure of worlds within worlds, with galaxies spread out in infinite distances."[185] This model, unfortunately of a piece with mere enumeration, is nonetheless an improvement in that it provides a 3-dimensional framework for beginning a description of the real, infinite universe.

## The Struggle for INFINITY

In the interests of "healing the nineteenth century breach between science and religion,"[186] indeterminists succeeded in persuading scientists to accept their belief in **finity**. Although one may doubt the benefits for science, it is surely good for religion: "Since modern science is now committed to a view of the physical universe as finite, certainly in space and probably in time, the activity which this same science identifies with matter cannot be a self-created or ultimately self-dependent activity. The world of nature or physical world as a whole...must ultimately depend for its existence on something other than itself....in a word, modern science, after an experiment in materialism, has come back into line."[187]

This is nothing more than the ancient, discredited metaphysics in sense I: what is beyond physics is not more physics, but non-physics. Determinists and indeterminists alike have learned that there is always more to the universe than meets the eye. They differ only on whether it is something or nothing.

Systems philosophy may accept microscopic infinity, but by definition, it will never accept macroscopic infinity. To do so would negate the myopia upon which systems philosophy is based. To accept INFINITY is to accept all the other Assumptions of Science and to discard the anthropocentric view of the universe once and for all.

From the concept of indivisible matter held by the classical mechanists, to the concept of the immaterial void held by the spiritualists, the resistance to

---

[184] Steen, L.A. "A New Perspective on Infinity." *New Scientist* 80 (1978): 448-51.

[185] Ibid., p. 450.

[186] Collingwood, *The Idea of Nature*, p. 156.

[187] Ibid., p. 155.

INFINITY has been strong. The quest for **certainty** and its search for the ultimate answer to the reason for the existence of the universe, periodically calls a halt to the question begging. Therein lies its fatal error. Today curved space has replaced the celestial sphere. Timid minds still seek shelter from the godless specter of INFINITY. As always, it will be to no avail.

# CHAPTER 9

# THE NINTH ASSUMPTION OF SCIENCE: RELATIVISM

*All things have characteristics that make them similar to all other things as well as characteristics that make them dissimilar to all other things.*

Whether or not we admit it, all thinking involves the comparison of one thing with another.[188] We try to understand what we do not know by comparing it to what we do know. The world obliges us in this respect by presenting us with a never-ending series of material objects, no two of which are either completely dissimilar or completely similar. Whether it should do so at all times and in all places is, however, a matter of serious philosophical contention.

## RELATIVISM VERSUS ABSOLUTISM

While most people probably agree that comparisons are highly important, they tend to view the comparisons themselves in either of two ways: as absolute or relative. With the development of formal logic, the absolutist approach achieved early domination. As part of the search for **certainty,** absolutists devised two primary laws that illustrate their thinking:

The Law of Identity or Equality, $A = A$, that is, every concept is equal to itself
The Law of Contradiction or Inequality, $A \neq A$

What distinguishes relativists from absolutists is the degree of flexibility they exhibit in thinking about these laws. According to relativists, the concepts of perfect equality and perfect inequality are only idealizations useful for describing the

---

[188] Conger, *New Views of Evolution*, p. 9.

intervening reality. Relativists believe that, in the strictest sense, both laws are false. There are no absolute equalities or absolute inequalities. The only way to consider any two things as exactly alike would be to ignore differences. There are no strict identities because all matter is in constant motion; no thing can be what it was just a moment before. Absolutists, on the other hand, tend to view objects as internally static, that is, containing matter without motion in accord with the assumptions of **separability** and **finity**. Only in this way can they insist on the literal imposition of logical ideality upon reality.

**Absolutism**, the indeterministic alternative to RELATIVISM, is the belief that some things may be perfectly identical or completely different from other things. **Absolutism** is consupponible with **certainty** and that bulwark of classical mechanism, the notion of finite universal causality. Only if an object could be described completely by a finite number of unchanging characteristics could it be absolutely identical to another object with the same description. And only then could its interactions with other objects be predicted with perfect precision. Anyone who believes this is possible must be considered an absolutist. Ironically, by this criterion Albert Einstein was an absolutist, not a relativist.

The philosophical opposition between RELATIVISM and **absolutism** arose early in the development of formal logic. The Greek sophists attacked the rigidity of logic simply by finding contradictions and exceptions in the statements of the logicians. The sophists were correct when they said such things as: "A white horse is not a horse, because it is a particular horse and a horse is a general horse." A white horse becomes a horse in general, only if we ignore its color. In short, the concept of a general horse is an abstraction, an idealization we use to give a class a name, to think about it, and communicate our thoughts to others. Only if all horses were identical could formal logic be completely adequate for their description.

For all the deficiencies pointed out by the sophists, the idealizations of formal logic were inescapable starting points for thinking. Even the lowliest animal must distinguish between food and nonfood. Mental activity itself involves elements of both RELATIVISM and **absolutism**. Throughout their daily lives all sentient creatures find it necessary to **assume** what amounts to identities or equalities to provide some stable basis on which to act. As always, however, the problem for philosophy is to avoid confusing ideality with reality. For this we must use all the tools at our disposal—particularly language. For example, relativist terms such as "similarity" and "dissimilarity" more accurately reflect reality than absolutist terms such as "equality" and "inequality."

# THE SIMILARITY-DISSIMILARITY CONTINUUM

To understand RELATIVISM one must also understand the problem of classification. In a sense, classification is one of the most important activities in science because its resolution comes with what science does best: make the subjective objective. This process requires an essential ingredient: cooperation. One observer may view two objects as similar because both objects have the same height; another may view them as dissimilar because the objects have different weights. Obviously, no agreement can be reached about similarity-dissimilarity unless the observers agree to compare the same characteristics. Until this is accomplished for a finite set of measurable characteristics, a classification or comparison must remain subjective rather than objective.

## Reasoning by Analogy

An "analogy" is a comparison that emphasizes similarity over dissimilarity. All analogies stand or fall on the appropriateness of the selection of the characteristics that are deemed to be similar. Because there are no two characteristics that are **exactly** alike, all analogies are vulnerable to error. As theoretical constructs, analogies are used to predict that if many of the characteristics of two objects have a high degree of similarity, then some of their other characteristics will have a high degree of similarity also.

As implied before, analogy is crucial to the whole process of knowing. It is impossible for us to consider the unfamiliar without reference to the familiar. To understand a thing, we are forced to consider only a few of the infinite number of qualities that it possesses. We must reduce the complex to the simple and the unknown to the already known. When asked to describe a taste sensation we have not experienced before, we find it impossible to answer without drawing an analogy with some other food. Thus in describing the taste of frog legs, for example, we might say: "They taste like chicken with a bit of fish mixed in." How ever we put it, the description must have simpler elements than the thing described.

Obviously, if all things were alike, we would not need to make comparisons (we would not be here for that matter). The fact that we can, and must, make comparisons is dependent on the dissimilarities that eventually become sufficient to destroy any analogy. For example, one could say that a person is like an automobile; they both consist of atoms and are influenced by gravity. The statement would be true as far as it goes. Of course, one could as easily point out the dissimilarities between people and automobiles. After all, according to INFINITY, the number of similarities and dissimilarities is infinite.

An analogy, like an assumption, must lead to understanding and accurate prediction or it will be discarded as useless. Thus it is not uncommon for scientists to discard analogies that run counter to "common sense," or that predict an outcome with which they do not agree. Consequently, a particular analogy often finds acceptance only after the necessity for it becomes clear in the broad social context. Humans, for example, were not considered similar to other animals until the scientific and commercial advantages of the analogy outweighed the religious objections.

The method of thinking by analogy is certainly not an exclusive property of *Homo sapiens*. Other animals must also make comparisons to survive. Consider the plight of the starving predator whose favorite food suddenly becomes scarce. It has no choice but to select another species that, by analogy, may be nearly as suitable for sustaining its life. The predator must continually abstract sense data from its environment, classify things into food and nonfood categories, and test its judgment in the external world.

In the above example it was advantageous to consider *Homo sapiens* and the predator as similar beings. I looked for similarities to understand the thought process in general. It is only by drawing such analogies that we can discover the connection between things in the universe. Without a plentiful use of analogy, a unified worldview is impossible. Thus, in following tradition, anthropocentric indeterminists who deny that such unification is possible or desirable, tend to disparage analogies involving the comparison of humans with other animals. Through an **absolutism** of one sort or another they attempt to disconnect the human being from its history and its present surroundings.

Of course relativist and absolutist alike are free to object to the choice of characteristics and the conclusions reached by a particular analogy. Only an absolutist, however, would claim that an analogy is impossible. If one goes far enough in generalizing the bias against analogy, one ventures close to intolerance of thinking *per se*.

## Reasoning by Disparity

A "disparity" is a comparison that emphasizes dissimilarity over similarity. All disparities stand or fall on the appropriateness of the selection of characteristics that are deemed to be dissimilar. Because there are no two characteristics that are **completely** different, all disparities are vulnerable to error. Like analogies, disparities are approximations. We use the method of disparity to divide things into different physical or mental categories, while at the same time, we use analogy to place things into one of two or more categories. Disparity is the result of divergence; analogy the result of convergence.

Categories or classes necessarily must form to the degree that things are isolated from their surroundings; they must dissolve to the degree that things are nonisolated from their surroundings. Although the classification process is a result of mental activity, it reflects the actual differentiation and integration of matter as well. To achieve the closest correspondence between the two, we must consciously use both analogy and disparity in formulating the experiments for finding out which is applicable in a particular context.

## Similarity Analysis

As scientists it is extremely important to reach agreement regarding a particular comparison. We do this by identifying the significant qualities of the objects to be compared and by collecting data in an effort to quantify those qualities. In transforming the subjective into the objective we must traverse three levels of comparison. The first level is purely qualitative and the most subjective. The second level is an attempt to measure characteristics in an objective way and then to compare the two sets of measurements in a subjective way. The third level is an attempt to compare the measurements themselves in an objective way.

At the elementary level of comparison, mathematics is not used in an explicit manner. Having no knowledge of simple arithmetic, a good witness, for example, might identify a suspect in a lineup. The sophistication of the elementary method is dependent on the number and significance of the qualities considered. Whether a person is short or tall is significant, but obviously this is insufficient for distinguishing him or her from all other humans. The inclusion of other characteristics such as sex, body build, hair color, and skin color narrows the possibilities still further, until, with enough of them, the person is uniquely distinguished on a purely qualitative basis. We admit the deficiencies of this highly subjective method whenever we attempt to mathematize the comparison by offering numerical estimates of height, weight, and age, thereby moving on to the intermediate level of comparison, the actual taking of measurements.

The intermediate level of comparison has only one general requirement: the actual measurement of some of the characteristics of the two things to be compared. In addition to the number and significance of the characteristics measured, we are also concerned here with the degree of accuracy (closeness to the "true" value) and precision (repeatability of the measurement). If we are to compare the measurements of one thing as a whole with the measurements of another thing as a whole, we can do it in either of two ways. The simplest is a visual examination of the data during which we compare the two measurements of a particular characteristic and then turn our attention to the data for the next characteristic. After evaluating two columns of such data we get an overall impression concerning whether the two objects are similar or dissimilar. In this intermediate method, the process of measurement reflects a degree of scientific maturity, but the subsequent handling of the data does not. For this we require additional mathematics that will help to decrease the subjectivity inherent in visual examination.

The advanced level of comparison uses a mathematical or statistical method for comparing the data acquired by measurement. There are numerous and very sophisticated methods for using mathematics in comparing two sets of data. The one I am most familiar with also happens to be about the simplest. The gist of it is given as equation 9-1, a formula I devised decades ago for calculating a single value that reflects the degree of similarity between two objects having numerous measured characteristics[189]:

$$d(A,B) = \frac{\sum_{i=1}^{n} R_i}{n} \quad (9\text{-}1)$$

Where:
$R_i = X_{iA}/X_{iB}$ if $X_{iB} > X_{iA}$
$R_i = X_{iB}/X_{iA}$ if $X_{iA} > X_{iB}$
$X_{iA}$ = Measurement for characteristic i of object A
$n$ = Number of characteristics measured

There are many such formulas, but this one, the SIMAN (sī'mun) coefficient, has special advantages.[190] In brief, we compare two objects by dividing the measured value of a characteristic of one object by the measured value of the same characteristic of the other. The basic rule for the calculation is that, for each variable, the divisor must always be the larger of the two measurements.

The SIMAN coefficient nicely illustrates the nature of the similarity-dissimilarity continuum. "Perfect" similarity, that is, identity, would result in a SIMAN coefficient of 1, while "perfect" dissimilarity would result in a SIMAN coefficient of 0. Comparisons of real objects always result in SIMAN coefficients less than 1 and greater than 0. Even measurements of the same object taken at two different times do not give values equal to unity. Both the thing being measured and its surroundings, including the measuring device, are continually changing. This is why no real thing can produce absolute values. The so-called "absolutes" of the

---

[189] Borchardt, Glenn. "Geochemical Similarity Analysis." *Geocom Programs* 6 (1972): 1-31.
[190] ———. "The SIMAN Coefficient for Similarity Analysis." *Bulletin of the Classification Society* 3 (1974): 2-8.

idealist are purely imaginary. Pi, for example, can be **calculated** at least a million more decimal places than it can be **measured**. The diameter and circumference of a real circle fluctuates over time—only the imagined, "ideal" circle does not. When real things are being compared, SIMAN coefficients of 1 or 0 reflect mathematical rounding rather than the actual existence of ideal things.

The SIMAN coefficient for the comparison between a 6-foot tall person and a 5-foot tall person would be 0.833. Comparisons between very tall people and very short people result, of course, in very low coefficients. There is no limit to the number of characteristics that can be used for a comparison. For example, if the 6-foot person and the 5-foot person had weights of 200 and 250 pounds, respectively, a comparison would yield a SIMAN coefficient of $[(5/6) + (200/250)]/2 = 0.817$ **for these characteristics**. The conclusions to be drawn, if any, from such comparisons depend on the context, which in this case generally would be a third person. Then, we can give a definitive answer to the question: "Is person A more similar to person B or to person C in regard both to height and weight?" As long as agreement can be reached as to which measurable characteristics are to be compared, we can decide objectively as to which two of three or more people or objects are most similar.

In spite of all this, and because the number of characteristics is infinite, similarity analysis always has a degree of subjectivity. This enters at the beginning in the selection of the characteristics deemed significant as well as in the portrayal of their interactions. It also enters at the end when a decision must be made regarding the level of similarity acceptable for a satisfactory answer to the question being asked.

## Examples of the Application of RELATIVISM

### The Electron

Both in philosophy and in science, RELATIVISM has traditionally taken a back seat to **absolutism**. As mentioned, the atomists maintained that each atom was identical to all others. Classical mechanics essentially adopted the same viewpoint, leaving itself open to the criticism of those who, like G. W. Leibniz, asserted that there are no two things that are perfectly identical. It is ironic that even after the theory of relativity became popular, **absolutism** still had its defenders in physics. This kind of thinking was illustrated in the words of Max Planck: "In contradistinction to chemical atoms all electric atoms (electrons) are found to be uniform and to differ from one another only in their velocity."[191]

---

[191] Planck, *Where Is Science Going?* p. 49.

In 1957, David Bohm echoed Leibniz's assumption and showed the connection between RELATIVISM and INFINITY: "Because every kind of thing is defined only through an inexhaustible set of qualities each having a certain degree of relative autonomy, such a thing can and indeed must be unique; i.e. not completely identical with any other thing in the universe, however similar the two things may be."[192]

In regard to electrons, Bohm claimed that it is "always possible to suppose that distinctions between electrons can arise at deeper levels."[193] Both RELATIVISM and INFINITY lead to the rejection of the possibility of actual identities in nature; both continue to be substantiated by the accumulating evidence. The debate on the electron was ended when experiments showed that there are at least three different types.[194]

## The "Conservation" of Parity

Another test of RELATIVISM involved the rejection of the "conservation" of parity in quantum mechanics. In brief, parity implied that atomic nuclei oriented in a particular direction would emit beta particles with the same intensity as they would when oriented in the opposite direction. Experiments finally showed that emission was not identical in both directions.

There are two ways of interpreting this rejection of the "conservation" of parity. In the indeterministic view, set forth by Ernest Nagel,[195] it is considered as a generally ignored falsification of CONSERVATION. In the deterministic view, the so-called "conservation" of parity was actually a restatement of **absolutism**. Its author had forgotten that parity, like perfect identity and perfect equality, is merely an idealization. What the idealists who supported this erroneous application were attempting to conserve, in effect, was the **idea** that both sides of a thing could be identical. Because identities do not exist in nature, because each thing is in continuous motion and is not even identical to itself from moment to moment, the attempt failed.

# TO THINK IS TO COMPARE

How ever one looks at the similarity-dissimilarity continuum—whether with a similarity coefficient or with some less quantitative view—one finds that

---

[192] Bohm, *Causality and Chance in Modern Physics*, p. 157.

[193] Ibid., p. 157.

[194] Weisskopf, "Contemporary Frontiers in Physics," p. 243.

[195] Nagel, *The Structure of Science*, p. 66.

comparisons on that continuum are fundamental to the method of thinking itself. Mostly we try to find the similarities in things, analogies, which are attempts to discover the familiar in the unfamiliar. It is generally easier to see what is different after we have seen what is alike. This bias extends to the language. There is no adequate antonym for the word "analogy" although the word "disparity" seems the best available. One may try to draw analogies, ignoring certain dissimilarities, or one may point out disparities, ignoring certain similarities. Our thoughts are forced into either of these two modes, for it is impossible to compare two things in terms of a single characteristic that is considered both the same and different at the same time.

The Ninth Assumption of Science, RELATIVISM, concerns the comparisons that are the basis for all statements, scientific or otherwise. All comparisons lie on the similarity-dissimilarity continuum. In nature, there are no absolute equalities and no absolute inequalities. Moreover, there are no analogies or disparities that cannot be contested by someone with contrary motives. As a result, the comparisons that we make in science and in everyday life have a single criterion for validity: usefulness.

# CHAPTER 10

# THE TENTH ASSUMPTION OF SCIENCE:
# INTERCONNECTION

*All things are interconnected, that is, between any two objects exist other objects that transmit matter and motion.*

The word "universe" portrays a fundamental property of existence: INTERCONNECTION. Whenever we try to think of any particular thing as a unity, we must view its parts as being interconnected. But to consider something as a part we must focus on it alone, momentarily suspending attention to its surroundings. Our thoughts necessarily must travel from parts to wholes and back again,[196] first viewing a thing as isolated from its surroundings, and then viewing it as a part of them.

RELATIVISM encouraged us to look at this aspect of the world by looking for the characteristics that make an object dissimilar from its surroundings, and then by looking for the characteristics that make the object similar to its surroundings. After having shown the ways in which two parts of the universe were unrelated, we showed the ways in which they were related. After disconnecting the world conceptually, we put it back together again.

In a similar vein, the Tenth Assumption of Science, INTERCONNECTION, recognizes that the world, like the ways in which we can view it, has both a discontinuous and a continuous nature. It is obvious that the universe contains countless examples of more or less discrete material objects and that each object displays a continuity within itself, but less obvious are two somewhat more sophisticated observations:

1. That the discontinuity between the object and its surroundings is not absolute—each contains things common to both, and

---
[196] Rosnay, *The Macroscope*, p. 161.

2. That the continuity within the object is not absolute—each object contains concrete discontinuities within.

The universe nowhere contains either empty, discontinuous space or solid, continuous matter. The ideas of absolute discontinuity and absolute continuity are only that: ideas. As usual, the reality lies in between, a reality we nevertheless cannot express without those ideas. The philosophical choice we need to make is not between an assumption of continuity and an assumption of discontinuity, but between a deterministic assumption that includes both of those ideas and an indeterministic assumption that does not. Thus, if discontinuity and continuity are to be considered qualities of **every** portion of the universe, we will have a lot of explaining to do whenever we are confronted by a portion that at first seems to be describable by one or the other, but not both. Outer space is a good example. How can both qualities, discontinuity and continuity, apply to what is commonly envisioned by the uneducated as completely empty?

Even well known materialists have lapsed into a confused idealism on this subject, apparently for lack of a clear definition of "space": "Space is continuous in the sense that between any two arbitrarily selected spatial elements (large or small, near or remote) there must always be in reality an element that joins them into a single spatial extent; in other words, between the elements of spatial extent there is no absolute separateness or isolation."[197]

What is this supposed to mean? As I see it, an "element of spatial extent" can represent one of two possibilities: either it is something or it is nothing. If it is nothing, then it is indeed empty space and hardly could form a connector of any sort. If an "element of spatial extent" is something, then it must have matter within it and therefore it can be considered to be an object. As an object (something), rather than a nonobject (nothing) it must be capable of transmitting matter and the motion of matter between what would otherwise be at least temporarily isolated "elements of spatial extent." Thus is born INTERCONNECTION, the deterministic assumption that **all things are interconnected, that is, between any two objects exist other objects that transmit matter and motion.** Through this objective and materialistic means we reject **disconnection**, the opposing indeterministic assumption that **between any two objects there can exist solid, continuous matter or empty, discontinuous space.**

---

[197] Konstantinov, F.V., and others, eds. *The Fundamentals of Marxist-Leninist Philosophy.* Moscow: Progress, 1974, p. 94.

## DISCONNECTION THROUGH THE IDEA OF PERFECT CONTINUITY

INTERCONNECTION assumes that the continuous quality of the universe is produced by discrete objects that, above all, are in continuous motion relative to each other. It might be objected that, with every two objects having another interposed between them, an infinite progression would produce solid matter of infinite density. One might wonder why the universe is differentiated at all. Such a view, however, would amount to a self-contradiction, because INTERCONNECTION assumes, along with INFINITY, that matter is infinitely subdivisible and that there is no end to the interposition of objects. For the universe to be completely undifferentiated, all its "parts" would have to be identical—a contradiction of RELATIVISM, as well as of the word "parts." Being solidly "connected" would, in effect, amount to a **disconnection** because there would be nothing to connect. The word "connect" implies the existence of more than one thing.

By hypothesizing the existence of perfect continuity, classical mechanism tended to use **disconnection** to do what all scientists must do in one way or another: ignore part of the universe. A solid, matter-filled object ceases to be of interest to science because it contains no thing within it that can be studied. Through this derivation of **disconnection**, mechanists naturally were led to overemphasize the external interactions of their model.

## DISCONNECTION THROUGH THE IDEA OF PERFECT DISCONTINUITY

The other way of deriving **disconnection** from the idea of perfect discontinuity was, of course, also present during the reign of classical mechanism, but has achieved an even more important place in systems philosophy. Today, the insides of the things we study cannot be ignored as easily as in the days of primitive scientific instruments. The notion of the solid, matter-filled object has been shoved into an ever-tighter corner, while the space between objects is still construed by many as though it were perfectly empty. By the empiricist, positivist, and operationalist[198] standards from which systems philosophy evolved, space is to be regarded as "perfectly empty" until evidence to the contrary is demonstrated. This **disconnection** of the object from its surroundings places renewed

---

[198] Operationalists believe that all theoretical terms in science must be defined only by their procedures or operations.

emphasis on the object itself. Derived in this way, **disconnection** leads to the overemphasizing of the internal and the ignoring of the external.

As the prevailing scientific worldview, systems philosophy generally achieves the first step in science. It correctly distinguishes the object or system from the rest of the universe. It goes part way toward achieving the second step by studying the interrelations between the parts within the system and, at times, even attempting to relate the system to its surroundings. But in general, systems philosophy tends to assume **disconnection**, always failing to the degree that it refuses to recognize that the surroundings of the system are as important as the system itself.

In a moment of great optimism, David Bohm wrote: "The universal interconnection of things has long been so evident from empirical evidence that one can no longer even question it."[199] Similarly, Barry Commoner, one of the first to emphasize the importance of the environment, declared that in ecology, the most important law is: "Everything is connected to everything else."[200] It would seem that the belief in INTERCONNECTION would be commonplace, but, as I will show, this is unfortunately far from true. It is the special mission of indeterminists to point out that the connections between things seldom are as obvious as Bohm and Commoner imply. Furthermore, INTERCONNECTION may be a useful generalization, but it remains for us to show, in each specific instance, what the most important connections are.

In the spirit of positivism, the belief in **disconnection** asserts that what lies between any two objects can just as easily be considered nothing as something. If by interconnection we only mean that objects exist in the same universe—though at a distance from each other, then Bohm's optimism is well taken. But if it is to mean more than that, if we are to reject the positivistic view altogether, then we need to show that things do not simply exist in the same universe, but that their motions invariably influence the motions of other things.

Historically, the belief in **disconnection** precedes that of INTERCONNECTION. After all, we begin life by believing that the rest of the world can be disconnected from us with a flick of the eyelids. Only with experience do we overcome solipsism and the tendency to view the world in a discontinuous fashion. About midway through this development it is natural that we should become dualists, assuming interconnections for some parts of the universe and denying them for others. Not being able to see all the connections between things, we continue to harbor the suspicion that, in some cases, there are no connections. As dualists we

---

[199] Bohm, *Causality and Chance in Modern Physics*, p. 143.

[200] Commoner, Barry. *The Closing Circle: Nature, Man, and Technology*. New York: Knopf, 1971, p. 33.

may be satisfied with a disjointed "world view." It is only after experience produces a grander vision that we learn to translate "Weltanschauung" as one word.

We really cannot know for certain whether or not the universe is properly described by INTERCONNECTION. Like the belief in CAUSALITY, the attempt to see the world as a unity must rest to some degree on "faith." Although the second step in science is to discover interconnections, the motions, the causes of effects, we are not always able to do this. We inevitably run out of evidence for the proposition that "between any two objects exist other objects that transmit matter and motion." It is ironic that at one time the hypothesizing of things for which there was no direct evidence was pretty much left to indeterminists. Today, however, it is the determinist who believes that interconnecting objects must exist, while it is the indeterminist who more often believes that they do not.

As implied in the chapter on COMPLEMENTARITY, the belief in **disconnection** leads to the idolization of the system itself as the source of its own development. If a thing is not subject to interactions with other objects in its surroundings, then it would exist, like the solipsist: all alone in a universe supposedly of its own making. In so far as we distinguish between things, but fail to relate them to other things, we reveal a juvenile bias in favor of **disconnection**, the extreme of which was manifest in sentiments attributed to Leibniz: "In every created thing God implanted the law of its own individual being, so that each being in the world is independent of and develops independently of all other things, following only the law of its own individual destiny."[201]

Today's systems philosophy repeats the same basic error. It begins with the child's egocentrism, develops along with the bourgeois notion of individualism, and retreats finally, to the citadel of free will. The illusion is maintained only by mentally disconnecting oneself from the environment of the present as well as from the memory and evidence of the environment of the past. Solipsism, egocentrism, anthropocentrism, and systems philosophy are merely variations on a theme.

To overcome this delinquent heritage let us review some of the supposed "disconnections" on which it is founded.

# SEARCH FOR THE UNIVERSAL DISCONNECTION

Like the search for the ultimate particle and the edge of the universe, the search for a universal **disconnection** fails with each improvement in knowledge. What first promises to be an absolute separation between an object and its environment is later found to be only a relative separation. According to

---

[201] Quoted in Planck, *Where Is Science Going?* p. 119.

COMPLEMENTARITY there can be no such thing as a completely isolated system. All objects exist in reality between the two ideals of complete isolation and complete nonisolation. There is always a transition zone or interface containing elements of both the system and the environment, a fact even recognized occasionally by the systems theorist.[202]

Absolutists believe otherwise. For them, the system is the system and the environment is the environment. Even as they dismiss the generalization that all things and their surroundings must necessarily undergo dynamic interaction at all times, their own bodies demonstrate against **disconnection**.

We are mostly water. When this water is inside the body we consider it part of ourselves, but after it has been exhaled in the breath, we do not. At what point should we consider this water to be nonhuman? As it exits the mouth? As it leaves the surface of the lungs? Where does the human being end and the environment begin?

Our skin daily loses epidermal cells that are continually being replaced. While these cells are still alive there is little question of their being part of the body. Long after they expire, the old cells lie loosely upon the skin as so much dead weight—they are actually part of the environment. Other cells are more firmly attached to the skin, although they may have died only moments ago. Still others have only a few moments to live. When a cell is alive and firmly attached it is clearly part of the body; when it is dead and loosely attached it is clearly part of the environment. The more closely we examine the transition between these two states, the more we must rely on arbitrary definition to maintain the belief in **disconnection**.

Before the invention of the microscope and the discovery of the molecular nature of things, the case for **disconnection** was much stronger than it is now. For instance, as recently as 1870, chemists looked on the transition from one chemical phase to another as a support for **disconnection**. They thought of the boundary between phases as absolute. A liquid was a liquid and a vapor was a vapor. Transition states intermediate between liquid and vapor were considered theoretically impossible, and the lack of data tended to support this view. All this became untenable after it was shown that liquid actually was transformed into vapor through a series of relatively homogeneous gradations in which many of the properties of both liquid and vapor were present at the same time.[203] So many phenomena exhibit such transition states that now they are cause for a general principle in chemistry.

INTERCONNECTION is closely allied with CONSERVATION. A thing is transformed into another thing only as it gains or loses matter or motion to other things in its environment. If this exchange was not possible, then the only way for

---

[202] Weinberg, *An Introduction to General Systems Thinking*, p. 149.
[203] Lewis and Randall, *Thermodynamics*. 1923, p. 186.

things to come into being would be through miraculous **creation**: the making of the material out of the nonmaterial. Any denial of the system-environment relationship amounts to a reiteration of the belief in **creation** as well as of **disconnection**. As we have seen in the discussion of COMPLEMENTARITY, such remnants of our indeterministic heritage are still very much active in the discipline of thermodynamics. As always, the best use of a deterministic assumption is at the point where data are scarce and speculation is rampant. One place where this occurs is in the study of the submicroscopic.

## The Interquantic Interconnection

With quantum mechanics, physicists have arrived at what is generally regarded as the culmination of the search for a universal **disconnection**. The transitions between energy levels within atoms occur by means of quantum jumps from one state to another. In the conventional view, the "transitions between these states are therefore not through a continuum of intermediate states."[204] David Bohm had an alternate view of subatomic phenomena: "Between the stable frequencies of oscillations exist unstable regions, in which the system tends rapidly to move from one stable mode to another. If we suppose that these transitions are very rapid compared with processes taking place at the atomic level, then as far as purely atomic phenomena are concerned they may be regarded as effectively discontinuous. Nevertheless, at a deeper level, they are continuous."[205]

At the atomic level, we know that when an electron is knocked out of orbit by a photon, the electron picks up a full quantum of motion. It is not possible for electrons and photons, indeed any of the particles we know of, to exchange a partial quantum. And yet INTERCONNECTION and INFINITY demand that subquantic exchanges must occur somewhere in the subatomic hierarchy. To demonstrate these, we would have to discover yet another level of particulate phenomena that would correspond with such subquantic exchanges of motion. Thus we expect that the particles we may eventually find **within** electrons and photons would interact with each other at the subquantum level rather than at the quantum level. It is presently unclear how or if we will be able to detect such phenomena. Even if this is achieved someday, indeterminists still could point to the lack of evidence for a "subsubquantic" level in support of **disconnection**. The new breed of determinist, however, will continue to assume that matter is infinitely subdividable and that the exchange of motion between those infinitely

---

[204] Hawkins, "The Thermodynamics of Purpose," p. 107.
[205] Bohm, *Causality and Chance in Modern Physics*, p. 81.

subdividable particles is not restricted to the quantum. We have rejected Greek atomism in the study of matter; let us reject it in the study of motion.

## The Intergalactic Interconnection

At first look, outer space appears to be an indeterminist's paradise, an irrefutable contradiction of INTERCONNECTION. To the uninstrumented eye, the regions between the stars seem to contain nothing at all, just empty space. If there is anything "connecting" the astronomical objects, it certainly is not obvious. Nevertheless, there are reasons to believe that INTERCONNECTION holds here too.

We believe the connection between two things to be direct and certain when matter is seen to extend from one to the other, such as in the case of the wire between two utility poles. What we sometimes forget is that the matter between the poles is really not "solid" in an absolute sense—it consists of atoms that are mostly "empty space." Unless we wish to resurrect atomism, we must agree that this holds for all other objects in the universe as well. For a connection to occur between two objects, we merely require there to be something else between them. This something else need not be "solid" matter.

At one time, astronomers also thought that the regions between the stars were void of matter. This was not surprising, partly because their early instruments were incapable of detecting matter there, and partly because their belief in INTERCONNECTION was weak. Of course, with the improvement in instrumentation, astronomers found that the interstellar regions contain gas and dust that form at least a partial interconnection.[206] Between the galaxies, too, areas formerly thought to be empty are nothing of the kind. Evidence is now accumulating in favor of an intergalactic interconnection consisting of various types of matter, which, although not always resolved with the strongest telescopes, may be detected in other ways. Research concerning the sun, for example, reveals a continuous emission of high-speed particles, many of which eventually leave the galaxy to travel through the intergalactic regions as part of the universal interconnection. Even those who still support the ballistic theory of light[207] must admit that space is not empty when light is traveling through it.

There is one other way of thinking about the universal interconnection. Beginning with Aristotle's idea of the impossibility of a vacuum and ending with the

---

[206] Science News. "More and More Mass: Universe Is Closing." *Science News* 114 (1978): 198.

[207] The belief that light is a particle traveling through empty space.

notion of a "neutrino sea,"[208] theorists have advanced the notion of ether, a medium permeating all things. Like the determinism-indeterminism struggle itself, the ether concept has gone through alternating periods of acceptance and rejection,[209] with recent work on the fringes of physics once again providing typically unheralded support.[210]

At first, experiments on atmospheric pressure and the production of modest vacuums led to the view that universal space was absolutely empty. Then the notion of a universal medium returned when the discovery of the wave nature of light seemed to require a medium to complete the analogy with other types of wave motion. This view survived until about 1910 when the Michelson-Morley experiments and special relativity led to its widely acclaimed rejection.

The historical ambivalence toward the ether is particularly reflected in Einstein's work. He has gone on record as thinking it irrelevant (1905),[211] unnecessary (1907),[212] necessary (1922),[213] unnecessary once again,[214] and finally, immaterial (1961).[215] At the end, Einstein refused to admit that he had left us with "completely empty space." Instead, his space was filled with a mathematically derived "field," which only incidentally contained no matter and had

---

[208] Dudley, "Is There an Ether?"; Dudley, H.C. "The Neutrino Sea—Hypothesis or Reality?" *Industrial Research*, December 1977, 51-54; Ruderfer, M. "Neutrino Structure of the Ether." *Lettere Nuovo Cimento* 13 (1975): 9-13.

[209] Whittaker, Sir Edmund. *A History of the Theories of Aether and Electricity: The Classical Theories*. Vol. 1. New York: Harper Torchbooks, 1951; ———. *A History of the Theories of Aether and Electricity: The Modern Theories, 1900-1926*. Vol. 2. New York: Harper and Brothers, 1953.

[210] Munera, H.A. "Michelson-Morley Experiments Revisited: Systematic Errors, Consistency among Different Experiments, and Compatibility with Absolute Space." *Apeiron* 5, no. 1-2 (1998): 37-54; Galaev, Y.M. "The Measuring of Ether-Drift Velocity and Kinematic Ether Viscosity within Optical Waves Band (English Translation)." *Spacetime & Substance* 3, no. 5 (2002): 207-24.

[211] Einstein, Albert. "On the Electrodynamics of Moving Bodies." In *The Principle of Relativity*, edited by A. Einstein, H.A. Lorentz, H. Weyl and H. Minkowski, 37-65. New York: Dover, 1905 [1923].

[212] ———. "Die Planck'sche Theorie Der Strahlung Und Die Theorie Der Spezifischen Wa"Rme." *Annalen der Physik* 22 (1907): 180-90, 800.

[213] ———. *The Meaning of Relativity*. 2 ed. Princeton, NJ: Princeton University Press, 1922.

[214] ———. and Leopold Infeld. *The Evolution of Physics: The Growth of Ideas from Early Concepts to Relativity and Quanta*. New York: Simon and Schuster, 1938.

[215] ———. *Relativity: The Special and the General Theory*. New York: Crown, 1961.

no material properties at all. In the currently accepted theory, light is viewed as both matter and motion, and the ambivalence remains. The eventual return to the wave theory of light will require some kind of material medium that would at the same time be part of the intergalactic interconnection.

Always, the region between objects and outside of objects has been a source of mystery. Because matter in "empty space" could be detected only with difficulty, it was usually assigned a lesser importance than the matter of the objects themselves. This naturally supported the system-oriented view that, by definition, fails to recognize the surroundings of the object to be as important as the object itself. To discard the bias of systems philosophy we must accept INTERCONNECTION and in so doing we must discard the concept of space as empty.

In the past, what we have called empty space has always turned out to contain matter. Outer space, formerly thought to be empty, is really filled with all manner of particulate matter. Even if one does not favor a new kind of ether, one can no longer be assured that the intergalactic regions are void of matter, and thus are evidence for a universal disconnection.

## THE NECESSARY CONNECTION

The inclusion of INTERCONNECTION, RELATIVISM, and INFINITY in a set of assumptions necessarily makes the reasoning somewhat circular: each assumption must have a degree of commonality with each of the others. Consupponibility without INTERCONNECTION is a contradiction in terms (Fig. 1). This is why you will never find a concordant explanation of the fundamental assumptions underlying classical mechanics or systems philosophy. Without INTERCONNECTION logical consistency is forced to yield to the persistent indeterministic claim that a unified worldview is impossible.

There is good reason indeterminists often maintain that assumptions are unnecessary: when placed side by side, the indeterministic alternatives to the Ten Assumptions of Science are contradictory and nonsensical. They result in a logical jumble startling for its incoherence. Mercifully, the mental effects produced by the indeterministic alternatives can occur only in the heads of those who temporarily abandon **disconnection**, attempting to find the interconnections among their own assumptions. At the outset, such a venture is unpromising. One must prepare for gross confusion even to begin a description of it:

## THE COMPLEAT INDETERMINIST

One might suppose that in the fantastic world of the compleat indeterminist there are no causes and no effects; things happen for no reason at all or just by absolute chance, which is also no reason at all. Paradoxically, there is no UNCERTAINTY in this world; everything is certain. Complete and perfect answers to questions are known even though this world is completely acausal. It is not always well known where the answers come from—perhaps they mysteriously appear and disappear like the material things and the motions of matter created and destroyed either by the spirit outside or the imagination inside. In this world, reversibility is the watchword, for time flows in both directions; events are repeated in endless reverie. In this strange world, there exist objects that are perfectly motionless inside and out. Nothing really happens because contact between these isolated objects never occurs. Each of the objects is its own ultimate particle: there is nothing inside of it and nothing outside of it. Many of the objects, however, are completely identical and others are completely different from other objects, although this cannot be true either because the disconnection between all the objects is absolute. In short, the world of the perfect indeterminist is **logically ridiculous**. Words cannot express adequately the confusion that abounds when one attempts to apply the indeterministic alternatives to the Ten Assumptions of Science.

Of course, modern indeterminists dare not go that far. Instead, they broach just a little **disconnection**, just a little **acausality**, just a little this and that, to avoid the implications of determinism. Once the logical roots of these derivative ideas are laid bare we can see how they form impediments to theoretical and practical progress. The modern derivatives then become no less absurd than the extremes posed by Berkeley in his day.

## INTERCONNECTIONS AMONG THE ASSUMPTIONS

The Ten Assumptions of Science, on the other hand, form a web of interconnections that are themselves worthy of lengthy study. With limited space here, only a few of the major interrelationships can be mentioned. With MATERIALISM we assume that, even though we are sentient beings, we are part of something much larger. As material portions of a material universe, we cannot do otherwise than obey laws similar to those we try to ascribe to other things. With CAUSALITY we assume an unbroken causal nexus that, because of its infinite character, is only partially available to our understanding. According to UNCERTAINTY, we may discover the most significant causes for a particular effect, always improving on the description after interacting with the external world.

Although these descriptions cannot be given in terms other than of matter and the motion of matter, we assume with INSEPARABILITY that matter and motion comprise an inseparable reality. Motion without matter and matter without motion are impossible, and it makes no more sense to try to conceive of matter **as** motion or motion **as** matter. With CONSERVATION we propose no beginnings or endings for matter and the motion of matter, an idea reiterated with COMPLEMENTARITY, which states that the Second Law of Thermodynamics is a law describing divergence, while its complement is a law describing convergence. According to IRREVERSIBILITY, each movement of each portion of the universe is unique—no two combinations or dissolutions occur in the same way twice. Each object has a unique relationship to the rest of the universe at any moment. This characteristic of time is consistent with the assumption of INFINITY, that the universe is infinite in both the microscopic and the macroscopic directions. This, in turn, is consistent with RELATIVISM that states that the manifestations of matter nowhere appear identical in two different places or at two different times. Furthermore, we do not expect to find any two things that are **completely** dissimilar and therefore independent of the INTERCONNECTION that describes the universe.

# ASSUMPTIONS AND THE INFINITE UNIVERSE

In elaborating upon the Ten Assumptions of Science I have tried to concentrate on areas where disagreement most often occurs, to show the fundamental differences between the deterministic and indeterministic viewpoints. The existence of these disagreements proves that these statements are assumptions, that is, matters of opinion. To make predictions, we must formulate assumptions, because in an infinite universe there is no obvious starting point. Our formulation and choice of assumptions must, of necessity, be based on our individual experiences with the world. And while no two experiences, and thus no two views of the world could be identical, their similarities must produce assumptions reflecting our unavoidable interaction with the external world. The explication and refinement of these assumptions is a never-ending process.

# CONCLUSIONS

The idealist philosopher, R.G. Collingwood, rightly claimed that science is based on presuppositions. This is tantamount to saying that science is based on "faith" rather than "fact." Presuppositions, he said, are logical starting points unrecognized and unexamined by scientists as long as they get pleasing results. Presuppositions become assumptions just as soon as they are stated—a process likely to occur only when results are not so pleasing. For me, the image of the entire universe exploding from a mathematical singularity was the last straw! Not being pleased, I delved into the subject and brought forth the assumptive choices stated in this book.

Collingwood somewhat unwittingly insisted on consupponibility—the proposition that if you can assume one assumption within a constellation, you must be able to assume all the others as well. Unfortunately for his idealistic cause, this amounted to a clarion call for a new kind of determinism, one based on the assumption that the universe is both microcosmically and macrocosmically infinite. In the final analysis, the opposing assumption of finity does not permit the logical interconnections required for consupponibility. In developing these Ten Assumptions of Science I confronted numerous contradictions based on finity. Once I discarded finity, the logic fell neatly into place.

The outcome of this inquiry, nevertheless, must be regarded as radical by today's standards: the universe had no beginning and will have no ending; time is motion; light is motion; there are only three dimensions; there is an ether; there is a simple mechanical complement to the Second Law of Thermodynamics. And last, but not least, these neatly consupponible assumptions support the replacement of the Big Bang Theory by the infinite universe theory.

Once I could overcome the idealizations of my youth, the philosophical possibilities were endless. Heretofore, I had been asking the wrong question about the universe: Why is there something, rather than nothing? The answer is that **nothing**, like completely empty space, is only an **idea**, just like **solid matter** is only an **idea**. As shown time and again in our experiments, the reality exists between these two idealizations. It turns out that it is impossible for the universe **not** to exist—everywhere and for all time.

The implications of these ten assumptions are so profound that the challenge to the Big Bang Theory seems almost incidental. We no longer need suffer the indignities of non-Euclidean curved space, massless particles, matterless motion, and a Second Law of Thermodynamics without its complement. In applying these assumptions, one sees quite a different world than most of us have been taught to accept. Those who thought the parts of the universe were a meaningless jumble of objects out of control will be encouraged instead to see the order that is there. Those who thought the control could be exerted exclusively from within themselves or exclusively on themselves will be encouraged to achieve a balanced view free from solipsism and fatalism.

The philosophy of a society becomes increasingly deterministic whenever people find the excess baggage of indeterminism and ignorance too heavy to bear. As in specialized science, it is the deterministic point of view that prevails when survival is at stake. Science and philosophy make accelerated progress when rapidly changing material conditions force people to seek the greatest accord between ideas and reality. Long-held philosophies suited to humanity's early development are becoming less and less viable. The myopia characteristic of today's scientific worldview, systems philosophy, will be discarded as the "environment," previously neglected, becomes increasingly prominent as a factor in our survival.

The rate of global population growth has been in decline since 1989. The "population bomb" hypothesized by systems philosophy never materialized. The global demographic transition centered on that year marks the unprecedented midpoint in our physical growth as a species. As we reach our ecological "carrying capacity" of 10 billion people, we will devise a scientific worldview that strives to achieve a theoretical balance in our consideration of the insides and the outsides of every single portion of the universe. Systems philosophy will be discarded as microcosmic and mechanism will be discarded as macrocosmic. Only a unification of the two will be adequate for humanity's new status as a species in tune with its surroundings.

# INDEX

Absolute chance, 23-24, 36, 41, 48, 117
Absolute truths, 33
**Absolutism**, 4, 10, 98-99, 101, 104-105
Abstraction, 18, 59, 80, 99
**Acausality**, 10, 22-24, 32, 36-37, 40, 84-85, 117
**Acausality**, definition of, 21
Acceleration, 48
Analogy, 49, 100-102, 106, 115
Anaximander, 61
Anthropocentrism, 111
*Argumentum ad ignorantiam*, 23, 85
Aristotelianism, 23, 30
Aristotle, 2, 23-24, 36, 41, 114
Astrologers, 62
Atomism, 45, 62, 89, 114
Axioms, 4, 95
Barnett, R. Michael, 92
Berkeley, George, 16
Bible, 67
Biblical flood, 65
Big Bang Theory, 5, 13, 67, 91, 119-120
Biology, 57, 65-66, 81
Bohm, David, 26-27, 32-35, 40, 45, 61, 91, 105, 110, 113
Bohm, David, on CONSERVATION, 61
Borchardt, Glenn, 103
Bronowski, Jacob, 42
Bruno, Giordano, 89
Brunsden, D., 85
Bryen, S.D., 48
Büchner, Ludwig, 46

Caloric theory, 63, 81
Cassirer, Ernst, 35
Catastrophism, 65, 81
CAUSALITY, 10, 16, 21-35, 40, 44-45, 50, 61, 67, 80-81, 84-86, 90-91, 99, 105, 110-111, 113, 117
CAUSALITY and MATERIALISM, 10, 29
CAUSALITY, example of, 27
Celestial sphere, 89, 97
**Certainty**, 11, 26-27, 30-31, 33, 36, 40-43, 48, 82, 85, 97-99
Chance, 16, 23-27, 33-34, 36-43, 48, 61, 86, 91, 105, 110, 113, 117
Chance, absolute, 23
Chaos, 36, 69
Classical mechanics, 48-49, 73-74, 104, 116
Classical mechanism, 45, 47-49, 88-91, 93, 99, 109
Clausius, Rudolf, 81
Collingwood, R.G., 2, 34
Commoner, Barry, 110
COMPLEMENTARITY, 10, 68, 70-71, 78, 82-83, 90, 111-113, 118
Comte de Buffon, 66
Conger, G.P., 49, 80
CONSERVATION, 10, 61-69, 81-82, 86, 105, 112, 118
CONSERVATION, definition of, 61
Consupponibility, 6, 12, 22, 33, 116, 118-119
Continuity, 107-109

Contradiction, 2, 5-6, 9, 16, 19, 23, 51, 68, 70-71, 86, 98, 109, 114, 116
Convergence, 10, 57, 68, 74-78, 83, 95, 101, 118
Copenhagen interpretation, 35, 41, 75, 91
Copenhagen school, 33-34, 40-41
Cornforth, Maurice, 48
Cosmogony, 55, 77, 95
**Creation**, 10, 61-62, 64-67, 69, 82, 94, 113
Darwin, Charles, 81
De Broglie, Louis, 32
Democritus, 25, 44, 89-90, 92
Devil, 65
Dialectical materialism, 3
Dialectics, 47, 77
**Disconnection**, 10, 12, 19, 108-113, 116-117
Discontinuity, 107-109
Discovery of time, 80
Disorder, 52, 68-71, 73-78, 82
Disparity, 4, 101-102, 106
Dissimilarity, 6, 99-101, 103
Divergence, 10, 68, 74-78, 83, 101, 118
Dogmatism, 11, 13
Dudley, H.C., 43, 115
Eddington, Sir Arthur, 41
Egocentrism, 111
Einstein, Albert, 15, 115
Einstein, Albert, and finite universal causality, 24
Einstein, Albert, as an absolutist, 99
Electron, 50-52, 92-93, 104-105, 113
Empiricism, 2
Energy, conservation of, 63
Energy, definition of, 46
Engels, Frederick, 47
Entropy, 35, 51-52, 68-69, 71, 73-78, 82, 84
Epistemology, 17, 31, 94
Equality, 98-99, 105

Equilibrium, 73, 82, 84-85, 87
ESP, 62
Ether, 43, 74, 115-116, 119
Evolution, 9, 13, 25, 42, 49, 52, 62, 65-67, 80-82, 85, 88, 90, 92, 98, 115
Experimentalist, 18
Extrasensory perception, 62
Facts, 1
Faith, 1-3, 7-8, 15, 17-20, 65, 93, 111, 119
Falsification, 4, 105
Fast, J.D., 35
Fate, 36
Feuerbach, Ludwig, on gods, 20
Field theory, 115
Finite universal causality, 22, 24-26, 28, 32, 40, 45, 81, 99
Finite universal causality, definition of, 24
**Finity**, 10, 41, 46, 88-92, 94, 96, 99, 119
First cause, 8, 64, 67
First Law of Thermodynamics, 10, 63, 81
Fleming, Donald, 49
Force, 48, 66, 74
Fortune, 36
Fossils, 65, 81
Free will, 8, 20, 23, 31, 111
Functionalists, 57
Galaev, Y.M., 115
Galileo, 89-90
Gamow, George, 55, 94
Geology, 65, 81
Ghosts, 18, 44
God, 65, 67, 111
Gods, 20, 44, 62
Hallucinations, 19
Heat death, 69
Heat, as matter, 63
Hegel, Georg, 10, 47, 55
Heisenberg Uncertainty Principle, 26, 31-32, 36, 40-41, 75

Heisenberg, Werner
Heraclitus, 23, 50
Hobbes, Thomas, 44
Holbach, Baron d', 45
Hooke, Robert, 89-90
Hutton, James, 81
Idealism, 15, 18, 32, 95, 108
Identity, 98, 103, 105
Ignorance, 36-37, 39, 41-43, 62, 75, 93, 95, 120
**Immaterialism**, 10, 15-16, 18-19, 23, 35
Industrial Revolution, 45, 82
Inequality, 98-99
INFINITY, 5, 10, 26-28, 32-33, 41, 46-47, 88-97, 100, 105, 109, 113, 116, 118
INFINITY, testability of, 5
INSEPARABILITY, 10, 44-48, 50-51, 53-58, 60, 63, 67-68, 79, 86-87, 118
INSEPARABILITY, definition of, 44
INTERCONNECTION, 10, 11, 50, 107-114, 116, 118
IRREVERSIBILITY, 10, 78-81, 83-85, 87, 118
IRREVERSIBILITY, CAUSALITY And, 80
IRREVERSIBILITY, INSEPARABILITY And, 79
Isolation, 4-5, 18, 48, 58, 70, 72-74, 77, 82-83, 85, 87, 90, 94-95, 108, 112
Jeans, Sir J.H., 31
June Goodfield, 80
Kant, Immanuel, 66
Kenyon, D.H., 42
Kuhn, T.S., 8
Kurtz, Paul, 67
Kwok, D.W.Y., 49
Laplace, Pierre Simon, 25
Laplacian determinism, 26, 28, 32-34, 36, 45, 81, 91
Lavoisier, Antoine, 63

Leeuwenhoek, Anton van, 90
Leibniz, G.W., 104
Lewis, G.N., 52
Light, 53-54, 90, 93, 114-116, 119
Loeb, Jacques, 49
Logic, 4-5, 22, 24, 60, 98-99, 119
Luck, 36, 43
Lucretius, 61
Makridakis, Spyros, 69
Mass, 27-28, 43, 48, 53-54, 57-59, 114
Mass, definition of, 57
Massless particles, 7, 120
MATERIALISM, 3, 10, 15-19, 29, 32-33, 44, 46-49, 59, 63, 67, 80, 83, 94, 96, 117
MATERIALISM, definition of, 15, 17
Mathematics, 31, 36, 40-42, 94-95, 102-103
Matter, 6, 10, 16-19, 21, 29, 33, 36, 39-40, 44-64, 67-72, 74, 76-84, 86-87, 90-94, 96, 98-100, 102, 107-109, 111-119
Metaphysicians, 7
Michelson-Morley experiments, 115
Miller, Hugh, 65
Minkowski, H., 87
Miracles, 62
Momentum, definition of, 58
Monkeys and typewriters, 41
Motion, conservation of, 63
Muehry, Henry, 92
Munera, H.A., 115
Mysticism, 22, 62
Nagel, Ernest, 57
Naïve realism, 16-17, 19, 46, 50
Nebular hypothesis, 25, 66
Negative entropy, 71
Negentropy, 71, 77
Neovitalism, 70, 73
Neutrinos, 93, 115
Newton, Isaac, 45

Newton's First Law of Motion, 73-74
**Noncomplementarity**, 10, 68, 78
**Noncomplementarity**, definition of, 68
Nonisolation, 70, 72-73, 77, 95, 112
Nonlinearity, 73
Objectivity, 1, 17, 29, 76
Onsager, Lars, 72
Operationalism, 109
Order, 51, 54, 67-78, 82, 120
Origin, 5, 8, 24-25, 40, 42, 66-67, 81, 91
Ostwald, Wilhelm, 49
Parity, 105
Pauling, Linus, 24-25
Peace, 19
Peirce, C.S., 23
Perfection, 31-32, 52
Pi, 104
Pigliucci, Massimo, 66
Planck, Max, 1, 18, 32, 80, 84, 94-95
Plato, 95
Pliny the Elder, 30
Positivism, 2, 110
Premises, 4
Presuppositions, 3-8, 13, 91, 119
Presuppositions, definition of, 4
Prigogine, Ilya, 69
Probabilistic models, 37
Probability, 24, 26, 37, 42, 86
Processes, 18, 65, 72, 79-80, 87, 94
Proctor, W.G., 53
Psychic healing, 62
Psychics, 62
Ptolemaic cosmology, 89
Quantum mechanics, 35-36, 71, 105, 113
Quark, 92
Quinn, Helen R., 92
Randall, Merle, 52
Reactionaries, 80
Realism, 17, 19
Reichenbach, Hans, 21

RELATIVISM, 4, 10, 11, 28, 98-100, 104-107, 109, 116, 118
**Reversibility**, 10, 48, 80-87, 94, 117
Revisionism, 11
Robinson, Abraham, 95
Rock of Gibraltar, 50
Rosnay, Joel de, 48
Ross, Hugh, 67
Russell, Bertrand, 22
Russell, Bertrand, on certainty, 31
Sagan, Carl, 42
Santayana, George, 79
Scheck, Florian, 36
Schroedinger, Erwin, 69
Scopes trial, 66
Scott, E.C, 62
Second Law of Thermodynamics, 10, 68-69, 73, 75, 78, 81, 118-120
Sediments, 65
Self-organizing, 73
**Separability**, 10, 44-45, 60, 82, 85-86, 99
Shannon, Claude, 75
SIMAN coefficient, 103-104
Similarity, 6, 24, 99-105
SLT, definition of, 68
SLT-Order Paradox, 68-73, 77-78
Sociology, 57, 81
Solar system, origin of, 25, 66, 81
Solipsism, 9, 110-111, 120
Sophism, 99
Souls, 44
Space, 58
Spacetime, 59-60, 70, 72, 91
Space-time, 6, 58
Specific causality, definition of, 22
Spinoza, Baruch, 44
Spirits, 18
Stasis, 21
Statistics, 37
Steinman, Gary, 42

Structuralists, 57
Subjective idealism, 15, 18
Subsystems, 73, 78, 85
System, 25, 37, 39, 48, 52, 55, 63, 66-78, 80-85, 87, 90, 110-113
System, closed, 70
System, isolated, 63, 68-70, 72, 74, 77, 82, 112
System, open, 70
Systems philosophy, 72-73, 80, 82-85, 88-91, 93, 95-96, 109-111, 116, 120
Ten Assumptions of Science, in brief, 10
Testability, 4
Theoretician, 18
Thermodynamics, nonequilibrium, 72
Things, 5, 8, 11, 13, 16, 18, 21, 25, 31, 35, 39, 44-48, 56-59, 61-62, 65, 72, 77, 80, 82, 92, 95, 98-102, 104-115, 117-118
Thomson, J.J., 92
Thornes, J.B., 85
Time independence, 86-87
Time, definition of, 59
Toulmin, Stephen, 80
Turchin, V.F., 3
UNCERTAINTY, 10, 26-27, 30-37, 39-43, 75, 84, 90-91, 95, 117
Uncertainty Principle, 26, 31-32, 36, 40-41, 75
UNCERTAINTY, CAUSALITY and, 30-31
Uniformitarianism, 65, 81
Universal causality, 22-29, 32, 35, 40, 45, 81, 99
Universal time, 59, 86
Velocity, 35, 48, 53-54, 58, 115
Velocity, definition of, 58
Vitalism, 70-71
War, 19
Wheeler, John, 16
Whyte, L.L., 52
Witness, 102
Zuckerkandl, Emile, 24-25

0-595-66263-3